Logic PhDs

Volume 2

The Algebra of Intensional Logics

Volume 1
Foundations of Combinatory Logic: Grundlagen der kombinatorischen
Logik. H. B. Curry, translated and presented by Fairouz Kamareddine
and Jonathan Seldin

Volume 2
The Algebra of Intensional Logics. J. Michael Dunn, with an
introductory essay by Katalin Bimbó

Series Editor

Jean-Yves Beziau   jyb@jyb-logic.org

The Algebra of Intensional Logics

J. Michael Dunn

with an introductory essay by
Katalin Bimbó

ISBN 978-1-84890-318-0

Published by College Publications

http://www.collegepublications.co.uk

Cover design by Laraine Welch

On the Algebraization of Relevance Logics. (A Preamble to J. Michael Dunn's PhD Dissertation)

Katalin Bimbó

1. Logic and Algebra in the Mid 1960s

Modern logic has its *algebraic roots* in the 19th century in the works of A. De Morgan and G. Boole. The algebraic approach to logic has been very successful, and the algebras that now bear the name of Boole turned out to be a useful abstraction beyond two-valued propositional logic. A finer analysis of sentences, however, requires more complicated algebras. The algebraization of quantification has been a recalcitrant problem for the better part of the 20th century, eventually, yielding polyadic algebras, cylindric algebras and complete algebras as some of the alternatives.

G. Birkhoff's work in *lattice theory* and A. Lindenbaum's approach to extricating an (interesting) *algebra from a logic* prepared the ground for a wholesome development of algebraic logic. The latter research area has been flourishing since the middle of the 20th century, to a considerable extent, thanks to the work of Polish logicians. In his dissertation, Dunn cites papers by McKinsey (some joint with Tarski), which deal with "non-classical" logics such as modal logics and intuitionistic logics. (It is ironic that while Tarski is a co-author on these papers, he remained an unwavering advocate of so-called "classical" logic—that excludes even normal modal logics—until the end of his life.) Two Polish logicians, H. Rasiowa and R. Sikorski followed up on the McKinsey–Tarski approach in their influential monograph of 1963 (see [45]). They gave a unified treatment of two-valued logic and intuitionistic logic, and emphasized an extended algebraic view, which included the use of topological and metric spaces as well as complete algebras. Certain modal and negation-free logics were also considered by them. However, they did not venture into the area of many-valued logic or relevance logics, despite [7], a paper by Białynicki-Birula and Rasiowa on quasi-Boolean algebras (i.e., bounded De Morgan lattices). It may be interesting to note that Rasiowa and Sikorski's interest with respect to algebraization is along the lines of considering the *Lindenbaum algebra* of a logic. That is, they focus on the set of *theorems of a logic* without non-logical axioms, although

1

they recall several axiomatic theories such as the theory of ordered sets, lattices, etc.

Dunn's dissertation goes beyond the repertoire of logics that have been algebraized by the 1960s. Relevant implication is not a residual of conjunction—unlike material implication or intuitionistic implication. Neither is relevant implication a monotone operation (in all its argument places)—unlike the usual modal connectives ($\Box$ and $\Diamond$). To put it quickly, the presence of $\rightarrow$ creates a markedly different algebra than Boolean or Heyting algebras, or even modal algebras. On the other hand, the distributivity of conjunction and disjunction has not been affected by concerns about viewing the conditional connective as implication. Ackermann's Π' calculus and the logic of entailment E (as well as several other relevance logics) share the fragment called first-degree entailments. Belnap and Spencer in [6] isolated the structures that they called intensionally complemented distributive lattices with truth filters (icdl's w/t-f, for short), which are special De Morgan lattices (with no fixed point for negation) with a subset of the carrier set selected. Dunn proved algebraic completeness theorems for first-degree entailments, using the four-element De Morgan lattice D and the eight-element icdl M_0. Interestingly, M_0's truth filter is not needed, and D has no filter that could be designated as a truth filter.

The *algebraization of R*, in Chapter X, which is perhaps, the most impactful achievement in the thesis, required the realization of the importance of t. Two-valued propositional logic, essentially, has one theorem, which is often taken to be excluded middle, $A \vee \neg A$. This formula ascended to a new height in fame with the advancement of intuitionistic logic, which does not allot the status of a theorem to its look-alike formula. Relevance logics are more versatile than two-valued logic, and they have many theorems, which are not all equivalent. A locally least theorem can be emulated as a conjunction of self-implications formed from a finite set of propositional variables (the locally relevant ones). However, there is no least theorem for every context unless t is included into the language of R. Dunn proved that the latter is necessary for the algebraization of R. (We shall say a bit more about the usefulness of t in the next section.)

For the sake of comparison, we should mention that there is another notion of algebraization for systems of logic in the literature. Blok and Pigozzi in [17] start with the usual idea of a congruence relation on the word algebra of a logic, which is defined for a pair of formulas A and B, by $A \rightarrow B$ and $B \rightarrow A$ being theorems, (where $\rightarrow$ is an implication-like connective in the language of the logic). But their focus moves from the algebra formed by equivalence classes of formulas in which there is a distinguished subset (the theorems) to the algebraization of consequence relations induced by axiomatic systems. The focus on deductive systems can be appreciated by noting that the modal logic S5 *is* or *is not algebraizable* depending on the details of how the axiomatic calculus is formulated (see [17, §5.2.1]). This approach finds intricate distinctions between axiomatic systems and directs attention toward theories and consequence relations. But mathematical theories formalized in the framework of two-valued logic typically involve quantification, which does not have a nice algebraic counterpart.[1] Relevance logics are often specified by an axiomatic calculus; however, a consequence relation defined via the commonly used (and lax) notion of proof is patently not relevant. Investigations related to the Lindenbaum algebra seem to have turned out more fruitful in the area of relevance logics.

Sections 2 elaborates on a crucial discovery in Dunn's dissertation and how it influenced later research in relevance logic. Section 3 takes a more encompassing view of the themes in the thesis and how those themes unfolded and developed in the work of Dunn later on. Section 4 gives a brief overview of the dissertation. Lastly, sections 5 and 6 record data pertaining to publications based on the dissertation and of the dissertation itself.

2. Intensional Truth and Intensional Conjunction

Dunn proved in Theorem X.3.1 that the Lindenbaum algebra of R is *not free* in the class of De Morgan semi-groups. He added t and also utilized the ∘ connective (fusion) in the definition of what he called *De Morgan monoids*. In the logic R, ∘ is a semi-group operation which is definable from negation and implication. (Cf. Theorem X.2.1.) We use the term "fusion" for ∘, because in T and E, it lacks associativity. In this section, we will quickly illustrate that these additions to R, and to other members of the relevance logic family, are ingenious in ways that go beyond the algebraization of relevance logic.

First, we will look at *Lindenbaum algebras* of some relevance logics. Second, we will mention some *sequent calculi*. Third, we recall some facts about the set-theoretic semantics for implicational relevance logics.

2.1. The relevance logics T, E and R

Nowadays, there are many relevance logics in the literature. Still, there are three (or four) that stand out, namely, T, E and R (and perhaps, B).[2] Axiomatizations of T, E and R may be found in [4, pp. xxiii–xxv], hence, we will not repeat them here. The way these logics can be thought to

---

[1]Cylindric algebras have to be locally finite, whereas polyadic and complete algebras are more than algebras from the algebraic purist's point of view.

[2]We mean the B from [42], not the logic of first-degree entailments, that is denoted by B in Dunn's dissertation. Dunn did not consider T, which is an interesting logic (in our opinion), but enjoyed much less popularity in the 1960s than E or even R did.

differ depends on the axiomatic formulation. However, with reference to the axiomatizations we mentioned, E extends T with specialized assertion (i.e., $((A \to A) \to B) \to B$) and modal distribution (i.e., $(((A \to A) \to A) \wedge ((B \to B) \to B)) \to (((A \wedge B) \to (A \wedge B)) \to (A \wedge B)))$, and further, R includes assertion (i.e., $A \to ((A \to B) \to B)$). t and $\circ$ can be conservatively added to each logic by two pairs of rules, which are $A \to (B \to C)$ implies $(A \circ B) \to C$, and vice versa; $t \to A$ implies A and vice versa. E also requires the t-specific axiom $(t \to A) \to A$ to be added. We will denote the extensions with $\circ$ and t of one of these relevance logics by attaching a superscript to the label of the logic—as in $T^{\circ t}$.

Definition 1. A *De Morgan lattice* is an algebra $(A, \wedge, N)$ that satisfies the equations (a1)–(a2).

   (a1) $a = a \wedge N(Na \wedge Nb)$

   (a2) $a \wedge N(Nb \wedge Nc) = N(N(c \wedge a) \wedge N(b \wedge a))$

The other lattice operation $\vee$, join is defined via a usual De Morgan law, that is, as $a \vee b =_{df} N(Na \wedge Nb)$. Then, (a1) is easily seen to be an absorption law $a = a \wedge (a \vee b)$, and (a2) is a permuted variant of distributivity, namely, $a \wedge (b \vee c) = (c \wedge a) \vee (b \wedge a)$. The latter two equations, which define distributive lattices in a concise way, come from [46], whereas the equations in Definition 1 are from [39].

Definition 2. A $T^{\circ t}$-*algebra* is an algebra $(A, \vee, N, :, \circ, e)$, where (a3)–(a10) hold.

   (a3) A is a De Morgan lattice with $\vee$ and N;

   (a4) A is an l-groupoid, (i.e., $a \circ (b \vee c) = (a \circ b) \vee (a \circ c)$);[3]

   (a5) $:$ is a residual of $\circ$, namely, $a \circ b \leq c$ iff $a \leq c : b$;

   (a6) e is (full) left identity for $\circ$, that is, $e \circ a = a$;

   (a7) N and $\circ$ interact according to $a \circ b \leq c$ iff $Nc \circ a \leq Nb$;

   (a8) $(a \circ b) \circ c \leq a \circ (b \circ c)$;

   (a9) $(a \circ b) \circ c \leq b \circ (a \circ c)$;

  (a10) $a \circ b \leq (a \circ b) \circ b$.

We may note that $:$ is definable from N and $\circ$ as $b : a =_{df} N(a \circ Nb)$ and detachment can be emulated utilizing e. Cf. Theorem X.2.1 and page 88 (the discussion following Theorem X.3.1) in Dunn's dissertation for the proofs of these claims for De Morgan semi-groups and De Morgan monoids.

---

[3]It may be expected that $(b \vee c) \circ a = (b \circ a) \vee (c \circ a)$ holds too. Indeed, this follows from (a5).

Lemma 1. *In a T^{ot}-algebra $b : a = N(a \circ Nb)$ holds, and if $a \leq b$ and $b : a \leq d : c$, then $c \leq d$.*

Proof. Consider the following series of iff's. $b : a \leq b : a$ iff $(b : a) \circ a \leq b$ iff $Nb \circ (b : a) \leq Na$ iff $a \circ Nb \leq N(b : a)$ iff $(b : a) \leq N(a \circ Nb)$. For the other direction of $\leq$, we proceed as follows: $a \circ Nb \leq a \circ Nb$ iff $Nb \circ N(a \circ Nb) \leq Na$ iff $N(a \circ Nb) \circ a \leq b$ iff $N(a \circ Nb) \leq b : a$.

From the two assumptions in the second claim, we get $e \leq b : a$ and $(b : a) \circ c \leq d$. But in an l-groupoid, $\circ$ is monotonic, hence, $e \circ c \leq (b : a) \circ c$, and so $e \circ c \leq d$, that is, $c \leq d$. $\qquad\square$

Furthermore, in the definition of a T^{ot}-algebra, (a8)–(a10) correspond to the prefixing $((A \to B) \to ((C \to A) \to (C \to B)))$, the suffixing $((A \to B) \to ((B \to C) \to (A \to C)))$ and the contraction $((A \to (A \to B)) \to (A \to B))$ axioms, that is, (A3), (A2) and (A4) in [4]. (The similarity of (a8)–(a10) to the axioms of the combinators B, B′ and W was remarked in [42].)

Certain properties of t—the least theorem—can be used to characterize the differences between the logics T^{ot}, E^{ot} and R^{ot} (or their Lindenbaum algebras) without adding any other (i.e., t-less) postulates, as we show in Lemma 2.[4]

Definition 3. *An E^{ot}-algebra is a T^{ot}-algebra in which (a11)–(a13) hold. An R^{ot}-algebra additionally satisfies (a14).*

(a11) $a : e \leq a$

(a12) $b : a \leq c : (c : (a : b))$

(a13) $(a : (a : a)) \wedge (b : (b : b)) \leq (a \wedge b) : ((a \wedge b) : (a \wedge b))$

(a14) $((c : b) : a) : ((c : a) : b)$

The labels for these algebras are intended to suggest that these are formulations of the Lindenbaum algebras of the respective relevance logics. And, of course, "R^{ot}-algebras" is just another name for De Morgan monoids that were introduced by Dunn. (We do not prove these claims here though.)

Lemma 2. *A T^{ot}-algebra in which $a \leq a \circ e$ holds is an E^{ot}-algebra. A T^{ot}-algebra in which $a = a \circ e$ holds is an R^{ot}-algebra.*

Proof. For a proof of (a11), we start with $(a : e) \circ e \leq a$, and note that e is upper right identity.

---

[4]We note that Meyer and Routley state in [42, §II] that on the background of B_+ (the positive fragment of the minimal relevance logic B mentioned above), the t-specific axiom, $(t \to A) \to A$ of E corresponds to $a \leq a \circ e$.

To show (a12), we note that ∘ is right monotonic in every T^{ot}-algebra, because ∘ is an l-groupoid operation. Similarly, it follows that if $a \leq b$ then $c : b \leq c : a$. We will use these inequations, as well as the transitivity of $\leq$ freely. Let us start with $(b : a) \circ a \leq b$. Then $e \circ ((b : a) \circ a) \leq b$ and by (a9), $((b : a) \circ e) \circ a \leq b$. By residuation, $(b : a) \circ e \leq b : a$. From $c : (b : a) \leq c : ((b : a) \circ e)$, by (a5) and (a9), $((b : a) \circ (c : (b : a))) \circ e \leq c$ follows. In two more steps, we have that $b : a \leq c : (c : (b : a))$, which is (a12).

(a13) may be written—using (the algebraic analog of) the necessity operation mentioned by Dunn in Chapter I—as $La \wedge Lb \leq L(a \wedge b)$. $a : (a : a) \leq a : e$ is immediate from e's properties and residuation. Then $La \wedge Lb \leq a : e$, hence, $(La \wedge Lb) \circ e \leq a$. Similarly, $(La \wedge Lb) \circ e \leq b$, by repeating the steps mutatis mutandis. Meet is glb (greatest lower bound), hence, $(La \wedge Lb) \circ e \leq a \wedge b$, from which we get $(a \wedge b) : (a \wedge b) \leq (a \wedge b) : ((La \wedge Lb) \circ e)$. Then, $((a \wedge b) : (a \wedge b)) \circ ((La \wedge Lb) \circ e) \leq a \wedge b$. And further, using that e is upper right identity, as well as, residuation we have $(a : (a : a)) \wedge (b : (b : b)) \leq (a \wedge b) : ((a \wedge b) : (a \wedge b))$.

Lastly, when e is lower right identity, $c : b \leq c : (b \circ e)$ is obvious. But from $(c : b) \circ (b \circ e) \leq c$, we get $(b \circ (c : b)) \circ e \leq c$ by (a9). Of course, e disappears, and by the right monotonicity of ∘, we obtain $b \circ (((c : b) : a) \circ a) \leq c$. Using (a9) one more time, $(((c : b) : a) \circ b) \circ a \leq c$, from where we get $(c : b) : a \leq (c : a) : b$, by residuation. $\square$

Dunn points out in Chapter X, section 6 that the algebraic equivalent of $(A \rightarrow ((B \rightarrow D) \rightarrow C)) \rightarrow ((B \rightarrow D) \rightarrow (A \rightarrow C))$, the restricted permutation axiom of E looks obscure. The inequation (a11) appears to be clear in comparison.

It is somewhat unfortunate that the algebras that emerged from logics other than two-valued logic and intuitionistic logic made their way slowly into the considerations of algebraists. For instance, [5]—compared to [45]—adds Post, Łukasiewicz and De Morgan lattices as special classes of distributive lattices, but Balbes and Dwinger appear to be unaware of a connection between De Morgan lattices and relevance logics (although they cite Belnap and Spencer's [6] on icdl's w/t-f). The idea, which is at the heart of Dunn's De Morgan monoids, namely, that having a fusion-like (or semi-group-like) operation together with its residual on the top of a distributive lattice yields an interesting algebra seems not have been fully realized in the algebraic literature for many more years after Dunn wrote his dissertation.

2.2. Sequent calculi and intensional truth

The least theorem or intensional truth, t turned out to be useful in the "Gentzenization" of relevance and other substructural logics. Dunn in [19] discovered that a careless application of the cut rule would spoil the intended formalization of R_+. His solution was to insist upon the insertion of t as a reminder that the cut formula was a theorem. He then proved that if $t \to A$ is a theorem of R_+, so is A.

In LK, which is Gentzen's original sequent calculus for two-valued logic, the step on the left-hand side (below) is unproblematic. The right-hand side cut is in Dunn's LR^+ calculus, and it illustrates the situation, when the left premise of the cut has nothing before $\vdash$, that is, A is a theorem.

$$\frac{\vdash \Gamma, A \qquad A, \Theta \vdash \Xi}{\Theta \vdash \Gamma, \Xi} \qquad\qquad \frac{\vdash A \qquad \alpha(A) \vdash B}{\alpha(t) \vdash B}$$

Γ, Θ and Ξ are finite (possibly, empty) sequences of formulas in the language of two-valued logic, whereas α is a structure composed of alternating extensional and intensional sequences of formulas in the language of LR^+. Further, $\alpha(A)$ indicates one or more occurrences of A inside α, which are replaced by t as indicated by $\alpha(t)$.

The use of t in sequent calculi, initiated by Dunn, led to formalizations of $T^t_\to$ and to the proof of the decidability of $T_\to$. However, there is a difference between $R_\to$ and $T_\to$. Fusion is a commutative semi-group operation (which is also upper idempotent) in R; hence, in $R_\to$, the exact place where a theorem has been cut out may be forgotten. Eventually, that theorem could have been moved to the left edge of the sequent before the cut. It is not so in $T_\to$, where the addition of t is essential in the formulation of $T_\to$ (really, of $T^t_\to$) as $LT^t_\to$.

Dunn's sequent calculus LR^+ (from 1969) is in [19]. (See also [3, §28.5] and [24].) An extension of LR^+ with necessity is in [4, §61]; sequent calculi for TW_+ and RW_+ are in [36] and [4, §67]. $LT^t_\to$ was introduced in [9]; t plays a crucial role not only in this sequent calculus, but it is essential to the decision procedure for $T_\to$. [14] and [15] present a detailed solution to the decidability problem of $T_\to$ using sequent calculuses. (See [10, §9.1] for a shorter description of the decision procedure.)

2.3.   Relational semantics for implicational fragments

Intensional truth, t and fusion are useful in the definition of set-theoretical semantics for relevance logics. Indeed, the natural way to define a semantics is to postulate a *binary operation* on some sort of theories or information sets. Each version of the set-theoretical semantics for relevance logics contained, at the time of its inception, a binary operation. (For a survey, see our [16].)

Of course, one might contend that the Routley–Meyer semantics can be formulated without t or $\circ$, because the set of theorems is not a situation anyway, and the analog of $\circ$ does not yield a situation from a pair of situations, in general. However, the inclusion of t makes it clear that the distinguished set of situations comprises the situations that include *all the theorems* of a logic, that is, all formulas A ẛt $t \rightarrow A$ is a theorem. Whereas the inclusion of $\circ$ makes transparent where the *ternary relation* comes from, which gives a more pleasing complexion to the whole semantics.

The situation changes if we want to provide semantics for the implicational fragment of, let us say, R. The straightforward generalization of the situations in the Routley–Meyer semantics are proto-theories the algebraic analog of which are *downward directed cones*. The set of theorems is easily seen to be such a cone once t is in the language. The inclusion of $\circ$, on the other hand, guarantees that suitable downward directed cones exist. (Cf. [13], [9] and [12].)

3. Impact and further developments

Dunn, certainly, continued his research in logic after writing his dissertation. We cannot attempt to follow up in detail on all the topics in his later work that have their root in the dissertation. Thus, we start with a reference to the *bibliography of his publications* in [11] (which is complete up to 2016), and we also recommend the papers in that volume that highlight aspects of Dunn's research in logic. We will use this brief section to point to some results by Dunn that have a "more algebraic flavor."

3.1. The third rule: γ

The logic of entailment, E was created by Anderson and Belnap through omitting the third rule, labeled γ, from Ackermann's Π' calculus (see [1, p. 119]).[5] This rule allows one to conclude B from A and $\overline{A} \vee B$, and may be conceptualized as "detachment for material implication." Another name for the rule is disjunctive syllogism, and it is central in Lewis's argument that justifies drawing an arbitrary conclusion from a contradiction, which is valid in two-valued logic.[6] Clearly, no relevance logic can allow γ as a rule that can be applied to any set of premises, because the consequence relation of the logic could not exclude irrelevant conclusions. It was hoped though—as stated in [2]—that the rule would be harmless when applied to theorems of E (and R, too), that is, that γ could be proved *admissible*.

Dunn considered equivalent algebraic formulations of γ in Chapter X, and he and Meyer proved the admissibility of γ in [40].[7] The first proof of

---

[5]By omitting Ackermann's *fourth* rule, labeled δ, Anderson created the logic of ticket entailment from E in 1960. In the usual axiomatizations of E, δ is replaced by an axiom such as restricted permutation or specialized assertion. See [3, §6, §8.1 and §8.3].

[6]See [3, §25.1] and [4, §80] on the disjunctive syllogism rule. Dunn presents in [24, §2.2] Lewis's proof from [38, pp. 248–251].

[7]There are further proofs of the admissibility of γ, for example, another one is in [3,

the admissibility of γ in [40] was *algebraic*, and a crucial step is to construct an E-algebra with no fixed point for negation given an E algebra that may have elements that are their own negation. The idea is to split elements of the latter kind and thereby normalize the E algebra. It is easy to state this, but there are many details to be worked out to ensure that the enlargement of the carrier set preserves the properties of the algebra that make it an E-algebra. Although Dunn and Meyer successfully applied their proof to some other relevance logics, most notably, to R and T (called P, in their paper), Dunn soon defined a group of logics in which γ is not admissible.

The logic R-mingle (RM) can be obtained from R by adding the axiom $A \to (A \to A)$, which is a special instance of the principal type of the combinator K. In [18], Dunn characterized denumerably many *pretabular extensions of* RM by giving their characteristic axioms. Then in Corollary 4, he delineated the extensions that admit γ and those that do not. The former group comprises consistent proper normal extensions of RM with characteristic matrices S_n versus the latter ones with S_{n+0}. To put it quickly, the difference between the characteristic Sugihara matrices is whether the elements are n positive and n negative integers, or those integers with 0 in between them. 0 is its own negation, but the "splitting trick" mentioned earlier cannot be used due to pretabularity.

3.2. More first-degree entailments

The logic of first-degree entailments (FDE) forms the stable core of several relevance logics. As we have already mentioned, its a common fragment of E, R and T.

De Morgan lattices are almost Boolean algebras, because quasi-complementation inherits the more pleasant properties of complementation, which provide self-duality, but omits the troublesome features. The availability of the De Morgan laws ensures that conjunction and disjunction are interdefinable, and with the double negation laws thrown in, normal form techniques can be used too. Indeed, it is obvious that if $A \to B$ is a formula in which A and B do not have occurrences of $\to$, then there is a formula $A' \to B'$ that is equivalent to $A \to B$ where A' is in disjunctive normal form and B' is in conjunctive normal form. Once the connectives of FDE are paired with the connectives of two-valued logic, it can be observed that $A' \to B'$ is an FDE theorem only if it is a two-valued tautology. Of course, the converse fails, which intrigued many scholars who went on to try to devise a test to sort out which two-valued tautologies are FDE theorems.

Dunn in [20] showed that *coupled trees* (a version of tableaux) for two-valued logic can be modified, indeed, be made more natural, so that it

---

§25.3]. A short historical account of proofs of γ is [49].

captures FDE. Coupled trees, however, do not reflect the idea that "a formula in DNF entails a formula in CNF." If the truth tree corresponding to B (in $A \to B$) is swapped for a falsity tree, then a connection emerges. In the two-valued truth-plus-falsity trees approach, closed branches are exempted from the covering requirement. Dunn [23] remedies this situation for FDE by stipulating that if p and $\bar{p}$ occur together on a branch, then p has to be relabeled into p' (a new variable) in every positive occurrence. Let $A' \to B'$ be as above. If there is a conjunction in A' that contains p and $\bar{p}$, then every literal p is changed to a fresh p' throughout the formula. Also, if there is disjunction in B' containing both q and $\bar{q}$, then every literal q is replaced by a new literal q'. Clearly, it takes finitely many steps to dis-identify the clashing variables. If the resulting formula is (still) a two-valued tautology, then $A' \to B'$ is an FDE theorem. Thus, eventually, it is possible *to sift out* the mischievous formulas from among all the two-valued tautologies, but it requires some effort not unlike actual sifting does.

The impact of Dunn's [20] is not limited to stimulating tableaux methods in relevance and substructural logics. Although many-valued logics had been around for half a century by the mid 1970s, they were often disparaged by some philosophers as mere technical devices. Dunn gave a new life to the four truth values that are used in the interpretations of first-degree formulas by recasting them as subsets of the set of two truth values $\{\,T, F\,\}$. Furthermore, he avoided the pitfall of deep metaphysical conundrums by considering truth values as *properties of information*. It is perfectly possible to have no information about p, in which case, p's truth value is $\emptyset$, and dually, one might learn both that q is true and q is false, hence, q's value is $\{\,T, F\,\}$. With a little relabeling, the four values turn into *neither* ($\emptyset$), *true* ($\{T\}$), *false* ($\{F\}$) and *both* ($\{T, F\}$), and the familiar truth functions interpret negation, conjunction and disjunction. (See [4, §81] for more details and applications.) This four-valued logic became to be known as the Dunn–Belnap or Belnap–Dunn logic, and it inspired generalizations into bilattices, trilattices and other many-valued logics.[8]

3.3. Algebraic semantics for other relevance logics

De Morgan monoids—introduced by Dunn in Chapter X—solve the problem of how to algebraize R. He also mentioned that E can be algebraized by a "brute force" method, namely, by turning a characteristic axiom of E (such as restricted permutation) into an equation. But the simplest instance of that axiom contains eight atomic formulas, and it is less than perspicuous.

---

[8]The history of the invention of this 4-valued logic can be found in [31]. Belnap's informal interpretation is detailed in [4, §81]. A sample of some of the latest publications related to the Dunn–Belnap logic includes [48], [43] and [47], as well as the forthcoming volume [44].

Entailment, E may have great reputation and impeccable philosophical underpinnings by combining (in a sense) modality and rigorous implication, but in hindsight, E is not the best behaved system from a technical viewpoint. Dunn, following a suggestion of S. MacColl, added the mingle axiom (i.e., $A \to (A \to A)$) to R as well as to E. R-mingle turned out to encapsulate some desirable features of relevance logics by retaining relevance between the antecedent and the consequent of a theorem—most of the times. But it also falters occasionally by violating variable sharing—a few other times.

However, R-mingle is all-around delightful with respect to its algebraic models. Sugihara matrices are chains, which makes them elegant structures. Dunn in [18] gave *algebraic completeness results for* RM (formulated with denumerably many propositional variables), and for its finite-variable fragments. RM itself does not have a finite characteristic matrix, however, it is pretabular, that is, all its proper normal extensions have a finite characteristic matrix—two-valued logic being the last such extension.[9]

De Morgan monoids can serve as models for R, moreover, they can provide *models for quantified* R, that is, RQ too. In Chapter VII, Dunn outlined a strategy to prove completeness for EQ (and RQ) via embedding the Lindenbaum algebra of the respective logic into a complete icdl which has further preservation and separation properties. However, as Dunn (with Meyer and Leblanc) shows in [41], there is a way to tweak the Lindenbaum construction for two-valued logic to guarantee that an RQ theory that does not prove A can be extended to a *prime* and *rich theory* that still does not prove A. (Richness means existential completeness together with its converse but stated as a condition for universally quantified formulas.) The construction is more complicated than the usual one, because care has to be taken to split elements that are fixed points for negation (the existence of which cannot be excluded based on a formula A not being deducible). Further, while a theory is being extended, certain formulas are being "kept out" (in a rejection set). The application of these ideas in [41] provides not only a completeness result for RQ, but also shows that γ remains admissible in the quantificational extension of R.[10]

3.4. Residuation and gaggle theory

The *idea of residuation* entered crucially into the definition of De Morgan monoids: $\to$ is the residual of $\circ$. In R, fusion has plenty of properties, which make it similar to operations like addition or multiplication. Notably, fusion is *not* the same operation as conjunction, because it does not

---

[9]Some results about RM by Dunn and by Meyer may be found in [3, §29.4 and §29.3].

[10]Using the insights from the case of RQ, Dunn and Meyer proved semantically the admissibility of the cut rule for Schütte's sequent calculus K_1 in [35].

satisfy idempotence. A binary operation with its residuals was considered by J. Lambek in the late 1950s to model the production of grammatical sentences (of the English language). Indeed, using the paradigm of an operation and its residuals, many substructural logics are algebraizable, with some connectives sticking together, so to speak.

Gaggle theory was introduced by Dunn in [25], and the label is a near acronym for "generalized Galois logics." Given an algebra with operations linked by (generalized) residuation on the background of a distributive lattice, the algebra has a relational representation in which an $n + 1$-place relation can be used to model every operation grouped together by residuation. This representation can be viewed as a *set-theoretic semantics* for the logic that has the algebra in question as its Lindenbaum algebra.

The construction of possible-world semantics goes back to the 1950s, and we will simply call this Kripke-style semantics here. For modal logics, the accessibility relation is binary, and for intuitionistic logic, the relation is binary again, despite the fact that implication is itself binary. Dunn in [21, 22] successfully used a binary relation to define a semantics for RM and RMQ. A difference from the modal and intuitionistic semantics of Kripke is that Dunn allows a situation to be *ambivalent* with respect to a formula. In other words, both T and F can be assigned to a formula. However, the binary relational approach seemed not to extend further to E, R and other relevance logics. [16] gives a historical outline of the invention of the ternary relational semantics and its close relatives by several logicians in the early 1970s, independently from each other. Although some people raised objections to the ternary relational approach—pointing at the lack of an easy informal reading of the relation—the relational approach proved to be a success.

Gaggle theory generalizes, in the first instance, the case of fusion to n-ary connectives, and to connectives (like intensional disjunction) which may have a different distribution pattern than ∘ has. In a series of papers [27, 26, 28, 29, 30], Dunn further developed the theory to make is applicable to the algebras of logics that do not contain a distributive lattice. Chapter 12 in [34] gathered the results in one place, and [13] gave a presentation using uniform notation and terminology.[11]

4. A Synopsis of the Thesis

This section provides a brief overview of the dissertation. Needless to say that this cannot replace reading the thesis itself.

---

[11] Research in the area of set-theoretical semantics for various logics—using relations or a combination of relations and operations—is flourishing. We will not attempt here to provide a list of references for this expansive literature.

The dissertation comprises *ten* chapters, which step by step move toward the *algebraization of* R, the logic of relevant implication and that of E, the logic of entailment. (An exception is Chapter IX, where an informal interpretation is presented for first-degree entailments.) Chapter X is the culmination of the thesis where *De Morgan monoids* are introduced.

Chapter I outlines how the two logics, R and E, came to be, and why they are in the focus of the thesis. Then, the chapter presents some elements of lattice theory together with the definitions of notions that are crucial for the treatment of logics. In particular, filters and ideals, as well as their prime versions are defined. Dunn dualizes Halmos's joke that logicians are duals of algebraists and he concludes that algebraists are duals of logicians. The theme of duality is more than a good-hearted pun here, because the core of the algebras of R and E are De Morgan lattices, which are self-dual with negation as an automorphism—just like Boolean algebras. Section 4 conveys the flavor of the approach that we might dub the Lindenbaum–Tarski-style algebraization of a logic. The last section generalizes intensionally complemented distributive lattices with a truth filter (abbreviated as icdl's w/t-f) to De Morgan lattices. When a truth filter is not required to exist, negation may have a fixed point.[12] In Boolean algebras, a fixed point for negation (which is complementation, $'$) trivializes the whole algebra; hence, the requirement of having more than one element in a Boolean algebra implies that there is no a s̃t $a' = a$. However, $a' \neq a$ (or $a \neq Na$, for De Morgan negation) is not an equation; it is the *negation* of an equation. This means that there is no obvious way to provide an equational characterization of De Morgan lattices in which N has no fixed point. Without an equational definition, Birkhoff's famous theorems about varieties are inapplicable. Indeed, Dunn uses homomorphisms (including isomorphisms) in the following chapters, what he can do because of the generalization of icdl's w/t-f to De Morgan lattices.

Chapter II explores the relationship between icdl's w/t-f and the eight-element lattice M_0. The latter can be visualized as the familiar eight-element Boolean algebra, but with complementation replaced by N (see p. 23). (Here M_0 is considered without any other operations such as $\rightarrow$ and $\circ$, which are often added to M_0.) N is "half-switched" compared to $'$; namely, complementation takes top to bottom, and co-atoms to atoms, and vice versa, whereas, N swaps two atoms in this configuration mapping two of the co-atoms to the atoms below them. In M_0, F_{+0}, the principal filter generated by $+0$ can be thought of as the truth filter. Dunn proves that

---

[12]An easy way to construct a De Morgan lattice with a fixed point for negation is by taking the divisor algebra of a square number. See [37] for the concept, and [8] for some concrete examples with Hasse diagrams.

every icdl w/t-f can be homomorphically mapped into M_0, with F_{+0} being the image of the truth filter in the icdl w/t-f. Furthermore, M_0 can support separation, that is, if $a \nleq b$, then there is an h st $h(a) \neq h(b)$. The next result is the embedding of any icdl w/t-f into a subalgebra of a product of M_0. Then, the results of the first two sections are generalized to complete icdl's w/t-f in sections 3 and 4. The existence of glb's (greatest lower bounds) and lub's (least upper bounds) for all subsets of the carrier set of the algebra is a property that cannot be expressed using finitary operations if there are infinitely many elements. However, this sense of completeness is interesting mathematically, for example, in the context of real numbers or power set algebras. Furthermore, complete algebras are useful if a logic has quantifiers; they also often emerge in the representation of algebras by sets. The last section provides two conditions, each of which is by itself necessary and sufficient for a complete De Morgan lattice to be a complete icdl.

Chapter III deals with De Morgan lattices, and in particular, the role of the 4-element De Morgan lattice D (see p. 37) as a homomorphic image of De Morgan lattices. Dunn proves that every prime filter of a De Morgan lattice determines a homomorphism into D, and conversely, the inverse image of F_p under each such homomorphism is a prime filter. Then he shows that every De Morgan lattice is isomorphic to a sublattice of a product of D, which parallels the theorem in the previous chapter about icdl's w/t-f and products of M_0. Lastly, the results are generalized to complete De Morgan lattices, by considering complete and completely prime filters, complete homomorphisms and completely join-irreducible elements. This chapter also provides some of the ingredients in the development of the parallel between **2**, M_0 and D with respect to their properties. The latter include the role that they play in the class of Boolean algebras, icdl's w/t-f and De Morgan lattices, respectively.

Chapter IV is entitled "Simplicity and normality." It is well-known that the 2-element Boolean algebra **2** is the only non-trivial simple Boolean algebra. Simplicity means that the only two congruences on the algebra are the total relation and the identity relation. Dunn shows that this notion straightforwardly generalizes to De Morgan lattices—by replacing Boolean complementation $'$ with intensional complementation N. Accordingly, "if $a \equiv b$ then $a' \equiv b'$" is swapped for "if $a \equiv b$ then $Na \equiv Nb$." In this sense, D and its sublattices are the only simple De Morgan lattices. Dunn introduces another notion of congruence that allows him to consider the analogous question for the class of icdl's w/t-f. The key idea is that a congruence should retain the distinction that is induced by the truth filter in addition to the usual properties a congruence has. That is, if $\equiv$ is a congruence (as usual), then for $\equiv$ to be a T-congruence, $a \equiv b$ cannot hold, if

$a \in T$, but $b \notin T$ (where T is the truth filter). Section 3 defines a notion of normality, which can be thought of as a De Morgan lattice being sufficiently versatile as De Morgan lattices go. Since a De Morgan lattice is a distributive lattice, normality characterizes a De Morgan lattice with respect to N (though the whole shape of the distributive lattice may be affected too). Now, instead of a truth filter T, a prime filter P should exist in a De Morgan lattice such that for some a, b and c, $a, \text{N}a \in P$, $b \in P$ but $\text{N}b \notin P$ and $c, \text{N}c \notin P$. Such a prime filter P exemplifies the three possible ways how an element and its negation can be related to P. Normal De Morgan lattices and icdl's w/t-f can be characterized, respectively, by having D and M_0 as their homomorphic images. For De Morgan lattices, Dunn also gives an elegant separation-like condition for normality: there are elements a and b s̃t, $a \wedge \text{N}a \nleq b \vee \text{N}b$. The interest in normality as defined here—rather than in normality as defined by Kalman—is that the Lindenbaum algebras of E and R are normal in the sense defined by Dunn.

Chapter V embeds icdl's w/t-f into ternary products of Boolean algebras, while *Chapter VI* embeds De Morgan lattices into binary products of Boolean algebras. Icdl's w/t-f have already been shown to be embeddable into a product of M_0. Now, the product is embedded into a 3-product of a direct product of **2**. This embedding connects the class of icdl's w/t-f to a very well understood class of algebras, namely, Boolean algebras. The latter have a representation theory due to Stone's work from the 1930s. Beyond the "prestige" that such a connection might provide, the link might open the door to new interpretations via transferring results.[13] The embedding of icdl's w/t-f yields a sub-3-product that Dunn calls special sub-3-product. The product is special, because the third component for every element is either all 1's or all 0's, which matches the mapping into F_{+0}, the truth filter in M_0. Thus, it is perhaps to be expected that omitting T, the truth filter from an icdl w/t-f, may lead to the binary representation of De Morgan lattices—as it is indeed shown to happen in Chapter VI. The representations have several equivalent formulations. Dunn is careful to point out that while the equivalences are effectively provable (i.e., their proofs do not require a use of the axiom of choice (AC) or its equivalents), the representations rely on Stone's theorem from 1937, which not only uses AC, but the use of AC appears to be unavoidable.

Chapter VII deals with a shared fragment of E and R, namely, with first degree entailments.[14] In view of the importance of De Morgan lattices

---

[13] We should point out that in the mid 1960s neither the Priestly representation of distributive lattices nor the Routley–Meyer semantics had been invented yet. Both date to the early 1970s.

[14] Several other relevance logics contain this fragment including the minimal relevance logic B and the logic of ticket entailment T. The latter logics are not mentioned in Dunn's

without fixed points for negation, it is unsurprising that the first theorem in this chapter establishes that the Lindenbaum algebras of E and R are icdl's. The argument mirrors the way relevance logics were created via the omission of problematic theorems (and correspondingly, axioms and rules). Two-valued logic has no fixed point for negation (the presence of which would lead to a disaster); hence, neither R nor E has any. Dunn proves—using algebraic methods—the completeness of the first degree fragment with respect to the class of icdl's, as well as M_0 and D. These are new proofs of earlier theorems by Anderson and Belnap, as well as by Smiley. The last section outlines what would need to be proved to show the completeness of quantified first degree entailments, that is, of first-order first degree entailments. Dunn uses the work of Rasiowa and Sikorski as a template for the algebraic treatment of quantification, and this approach relies on complete algebras (complete Boolean algebras, or for relevance logics, complete De Morgan lattices) with topologies. He describes step by step the difficulties that would have to be overcome to obtain a theorem that would parallel the two-valued quantificational case.[15]

Chapter VIII recalls an axiomatic formulation of the logic of first degree entailments (called B), and connects its Lindenbaum algebra to De Morgan lattices and icdl's. First of all, the Lindenbaum algebra of B is an icdl, hence, it is also a De Morgan lattice. Second, soundness of B holds with respect to the class of De Morgan lattices as well as icdl's. Third, the Lindenbaum algebra of B with n propositional variables is a free De Morgan lattice with n generators, and a free icdl with n generators. Fourth, a condition that characterizes a set of elements as free generators of an icdl can be transposed into the situation of De Morgan lattices. This gives that a free De Morgan lattice with n generators is a free distributive lattice with $2n$ generators. Hence, fifth, there is an upper bound on the number of elements of the free De Morgan lattice with n generators, namely, $2^{2^{2n}}$.

Chapter IX gives an intuitive interpretation for first degree entailments. It is natural to assume that sentences are about something, and this has been a recurring theme in philosophical thinking about reasoning carried out using language. In this chapter, Dunn analyzes certain ideas concerning aboutness that were championed by Nelson Goodman (a prominent analytic philosopher in the 20th century). Then he goes beyond Goodman's claims to clarify what sentences with negations talk about, introducing the idea

---

thesis. Although T was created by Anderson in 1960, it is "farther" from two-valued logic than R or E, and it often receives less attention than those two.

[15] We may note that the most popular algebraization of quantified two-valued logic seem to have gone into other direction than complete Boolean algebras. Cylindric algebras enrich a Boolean algebra with (possibly) infinitely many unary operations and constants (i.e., zero-ary operations).

that sentences can have positive or negative content about a given topic, but also no content at all. This evolved into the Dunn–Belnap 4-valued logic. De Morgan negation behaves dually with respect to conjunctions and disjunctions, hence, it can be used to explicate the topics that sentences connected with "and" and "or" are about. Then the largest subdirectly irreducible De Morgan lattice D can be viewed as capturing the four possible types of "aboutness situations" given just one topic.[16]

Chapter X is the pinnacle of the dissertation, because this is where Dunn algebraizes R, the logic of relevant implication, as a whole. A reduct of the Lindenbaum algebra of R, namely, the algebra of B had been dealt with in the previous chapters. However, now a commutative, square-increasing semi-group is superimposed on a De Morgan lattice giving what Dunn calls De Morgan lattice-ordered semi-groups. Relevant implication is, certainly, not a semi-group operation. But fusion, which is definable from negation and implication, is. Fusion is not always included in the language of R (or E), indeed, it does not occur in the axiomatization in Chapter I. However, Dunn proves that every De Morgan lattice-ordered semi-group is residuated, that is, it has a relevant implication. Moving into the opposite direction, from relevant implication and De Morgan negation, fusion is definable.

However, simply combining a De Morgan lattice and a commutative, square-increasing l-semi-group does not do justice to the algebra of R, in which implication and negation do interact. For example, the so-called reductio axiom and (all forms of) contraposition hold. Dunn postulates three pairs of quasi-inequations, which look similar to the residuation conditions, but they involve negation, as part of the definition of De Morgan lattice-ordered semi-groups.

Along the lines of the previous chapters, one would like to see the Lindenbaum algebra of R to be free in the class of De Morgan semi-groups. However, Dunn proves that this is not the case. R may be axiomatized taking as one of the axioms specialized assertion $((A \to A) \to B) \to B$. Assertion is $C \to ((C \to B) \to B)$, which specializes into the former formula by substituting $A \to A$ for both C's and detaching the first. In turn, assertion is specialized permutation, that is obtained from $(A \to (C \to B)) \to (C \to (A \to B))$ by detaching an instance of self-implication (i.e., $(C \to B) \to (C \to B)$) after substituting $C \to B$ for A. R was defined in Chapter I with specialized assertion as an axiom, however, Dunn pointed out that the specialized assertion axiom may be replaced by permutation (without altering the set of theorems). Now, he shows that similar maneuvers cannot be always performed in lattice-ordered De Morgan semi-groups. The counterexample is a four-element chain without an identity element for

---

[16]See [50] for a book-length treatment of aboutness.

∘, in which the algebraic version of specialized assertion does not hold, but the four-element algebra is an intensional semi-group.

The axiomatization of R did not contain the identity constant t, and Dunn proves that R does not have a definable identity constant, in general. Local versions of t are definable, and such local constants have been known in the literature previously. On the other hand, t can be introduced into the language of R—without adding to the stock of theorems in the old vocabulary; that is, t can be added conservatively. With two axioms included into an axiomatization of R, one can guarantee that t behaves as intended. The Lindenbaum algebra of R is isomorphically embeddable into the Lindenbaum algebra of R^t. Finally, R^t algebraizes into a *De Morgan monoid*, which is free in its class. The insight concerning the importance of the identity constant to support the algebraic analog of the detachment rule allows Dunn to prove theorems similar to those in the case of first-degree entailments. In particular, provability in R^t coincides with validity in De Morgan monoids.

The newly introduced class of algebras is put to use in section 5. The third rule in Ackermann's calculus Π' has been excluded both from R and E, by Anderson and Belnap. But they conjectured that this rule—called γ—may be admissible in E (and R), that is, γ may be applicable when the premises are theorems. Dunn considers several algebraic formulations of the rule, and shows that they are equivalent. Then he proves that, to start with, γ is not admissible in the class of De Morgan monoids. The counter example he gives relies on a fusion matrix for the six-element sub-icdl of M_0 that excludes -2 and $+2$.

The next section is concerned with the addition of a necessity operation to R (and R^t). Robert K. Meyer conjectured in his PhD thesis (1966) that E can be viewed as a modalized version of R. Dunn provides an algebraization of this extension, while he also compares the necessity in R^t with the necessity in S4. More precisely, he compares the properties of the closure operator (which is the analog of necessity) in R^t and in the work of McKinsey and Tarski.[17] The last section gives a brief overview of problems closed (in the dissertation) and problems remaining open (for the relevance logics E and R, including their fist-order versions).

---

[17]We should note that Meyer partially proved his own conjecture in his thesis, namely, for the implicational fragments of E and R. It was not unreasonable to expect, at the time Dunn was writing his dissertation, that E could be faithfully recast as R with necessity. It was not until seven years later that L. Maksimova proved that the equivalence between E and modal R failed in their positive fragments.

5. Previous Publications Based on Parts of the Thesis

The dissertation has been available in the form of microfiche from UMI (University Microfilms International) since 1966, and more recently, as a PDF (portable document format) file of a scan of the typewritten copy from ProQuest. Some parts of the thesis appeared in publications. First of all, (essentially) Chapter 2 has been published as Dunn and Belnap [33]. As the bibliography indicates, a talk based on this chapter was presented by Dunn at a meeting of the Association for Symbolic Logic in 1965. The abstract of the talk was published as [32].

The bibliography in the dissertation, lists *Entailment* as a forthcoming book. It was published in two volumes ([3] and [4]) some years later. Dunn is a contributor to the first volume and a co-author of the second. In the first volume, §18 and §28.2 are directly related to Dunn's dissertation. The time interval between 1966 and 1975 (i.e., Dunn's dissertation and the publication of volume I) allowed for some reshaping of the text, but the two sections mentioned are based on parts of Chapters I, III, VII, VIII and X of the thesis.[18]

The informal interpretation from Chapter IX, which chops propositions into a positive and a negative component, became the foundation for [20].[19] This is one of the most cited papers of Dunn, especially, in philosophical circles. The interpretation of the four truth values as combinations of "told to be true" and "told to be false" values freed the truth values from any metaphysical burden and inspired further ideas that were explored by philosophers. The title of the paper talks about intuitive semantics, which certainly, is in the paper. However, Dunn also develops "coupled trees" for first-degree entailments, which are a modification of tableaux for two-valued propositional logic. Arguably, these coupled trees are the most elegant proof system for first-degree entailments.

6. The LATEX'ed Version of the Dissertation

The dissertation has been typeset anew, using the OCR'd version of the PDF scan of the original. The typeset version—while faithful in content—is not intended to preserve the look of the original, which was produced on a typewriter. Indeed, the present book utilizes the capabilities of LATEX to create a reader-friendly text, which is perhaps, aesthetically more pleasing than the original typescript. To this end, some minor changes have been made. Firstly, the symbols are the proper symbols available in TEX. (E.g., Π' replaces II'.) Secondly, chapters, sections, and similar structural com-

---

[18]Some other sections in the first volume of *Entailment* written by Dunn are based on the continuation of the research in his dissertation.

[19]As Dunn says in [20, p. 166], "Basically it all stems from my dissertation ...."

ponents have been replaced by the ones that are provided in LaTeX. As a result, the page numbering here does not coincide with the page numbering in the typescript. To enable a reader to cite the dissertation with page numbers, the original page numbering is indicated (from Chapter I on) by small insertions such as $^{123}\|_{124}$, with the reading that page 123 ends and page 124 starts at this point. (Accordingly, the page numbers referenced in the "Appendix: Notation" have been replaced by the page numbers here. The reader can easily see the original page number by jumping to the page referenced.) Thirdly, we hope to have enhanced the readability of the dissertation by using fonts from the Computer Modern series, by correcting obvious misprints and by replacing underlining with italics. However, we should emphasize that no significant changes have been made to the thesis.

References

[1] Wilhelm Ackermann. Begründung einer strengen Implikation. *Journal of Symbolic Logic*, 21:113–128, 1956.

[2] Alan R. Anderson. Some open problems concerning the system *E* of entailment. *Acta Philosophica Fennica*, 16:9–18, 1963.

[3] Alan R. Anderson and Nuel D. Belnap. *Entailment: The Logic of Relevance and Necessity*, volume I. Princeton University Press, Princeton, NJ, 1975.

[4] Alan R. Anderson, Nuel D. Belnap, and J. Michael Dunn. *Entailment: The Logic of Relevance and Necessity*, volume II. Princeton University Press, Princeton, NJ, 1992.

[5] Raymond Balbes and Philip Dwinger. *Distributive Lattices*. University of Missouri Press, Columbia, MI, 1974.

[6] Nuel D. Belnap and Joel H. Spencer. Intensionally complemented distributive lattices. *Portugalie Mathematica*, 25:99–104, 1966.

[7] Andrzej Białynicki-Birula and Helena Rasiowa. On the representation of quasi-Boolean algebras. *Bulletin de l'Académie Polonaise des Sciences*, 5:259–261, 1957.

[8] Katalin Bimbó. Functorial duality for ortholattices and De Morgan lattices. *Logica Universalis*, 1:311–333, 2007.

[9] Katalin Bimbó. Relevance logics. In D. Jacquette, editor, *Philosophy of Logic*, volume 5 of *Handbook of the Philosophy of Science* (D. Gabbay, P. Thagard and J. Woods, eds.), pages 723–789. Elsevier (North-Holland), Amsterdam, 2007.

[10] Katalin Bimbó. *Proof Theory: Sequent Calculi and Related Formalisms*. CRC Press, Boca Raton, FL, 2015.

[11] Katalin Bimbó, editor. *J. Michael Dunn on Information Based Logics*, volume 8 of *Outstanding Contributions to Logic*. Springer Nature, Switzerland, 2016.

[12] Katalin Bimbó. Some relevance logics from the point of view of relational semantics. *Logic Journal of the IGPL*, 24(3):268–287, 2016. O. Arieli and A. Zamansky (eds.), *Israeli Workshop on Non-classical Logics and their Applications* (IsraLog 2014).

[13] Katalin Bimbó and J. Michael Dunn. *Generalized Galois Logics: Relational Semantics of Nonclassical Logical Calculi*, volume 188 of *CSLI Lecture Notes*. CSLI Publications, Stanford, CA, 2008.

[14] Katalin Bimbó and J. Michael Dunn. New consecution calculi for $R^t_\to$. *Notre Dame Journal of Formal Logic*, 53(4):491–509, 2012.

[15] Katalin Bimbó and J. Michael Dunn. On the decidability of implicational ticket entailment. *Journal of Symbolic Logic*, 78(1):214–236, 2013.

[16] Katalin Bimbó and J. Michael Dunn. The emergence of set-theoretical semantics for relevance logics around 1970. In K. Bimbó and J. M. Dunn, editors, *Proceedings of the Third Workshop, May 16–17, 2016, Edmonton, Canada*, volume 4(3) of *IFCoLog Journal of Logics and Their Applications*, pages 557–589, London, UK, 2017. College Publications.

[17] Willem J. Blok and Don Pigozzi. *Algebraizable Logics*. Number 396 in Memoirs of the American Mathematical Society. American Mathematical Society, Providence, RI, 1989.

[18] J. Michael Dunn. Algebraic completeness results for R-mingle and its extensions. *Journal of Symbolic Logic*, 35:1–13, 1970.

[19] J. Michael Dunn. A 'Gentzen system' for positive relevant implication, (abstract). *Journal of Symbolic Logic*, 38:356–357, 1973.

[20] J. Michael Dunn. Intuitive semantics for first-degree entailments and 'coupled trees'. *Philosophical Studies*, 29:149–168, 1976.

[21] J. Michael Dunn. A Kripke-style semantics for R-mingle using a binary accessibility relation. *Studia Logica*, 35:163–172, 1976.

[22] J. Michael Dunn. Quantification and RM. *Studia Logica*, 35:315–322, 1976.

[23] J. Michael Dunn. A sieve for entailments. *Journal of Philosophical Logic*, 9:41–57, 1980.

[24] J. Michael Dunn. Relevance logic and entailment. In D. Gabbay and F. Guenthner, editors, *Handbook of Philosophical Logic*, volume 3, pages 117–224. D. Reidel, Dordrecht, 1st edition, 1986.

[25] J. Michael Dunn. Gaggle theory: An abstraction of Galois connections and residuation, with applications to negation, implication, and various logical operators. In J. van Eijck, editor, *Logics in AI: European Workshop JELIA '90*, number 478 in Lecture Notes in Computer Science, pages 31–51. Springer, Berlin, 1991.

[26] J. Michael Dunn. Partial gaggles applied to logics with restricted structural rules. In K. Došen and P. Schroeder-Heister, editors, *Substructural Logics*, number 2 in Studies in Logic and Computation, pages 63–108. Clarendon, Oxford, UK, 1993.

[27] J. Michael Dunn. Star and perp: Two treatments of negation. *Philosophical Perspectives*, 7:331–357, 1993. (Language and Logic, 1993, J. E. Tomberlin (ed.)).

[28] J. Michael Dunn. Gaggle theory applied to intuitionistic, modal and relevance logics. In I. Max and W. Stelzner, editors, *Logik und Mathematik. Frege-Kolloquium Jena 1993*, pages 335–368. W. de Gruyter, Berlin, 1995.

[29] J. Michael Dunn. Generalized ortho negation. In H. Wansing, editor, *Negation: A Notion in Focus*, pages 3–26. W. de Gruyter, New York, NY, 1996.

[30] J. Michael Dunn. A representation of relation algebras using Routley–Meyer frames. In C. A. Anderson and M. Zelëny, editors, *Logic, Meaning and Computation. Essays in Memory of Alonzo Church*, pages 77–108. Kluwer, Dordrecht, 2001.

[31] J. Michael Dunn. Two, three, four, infinity: The path to the four-valued logic and beyond. In H. Omori and H. Wansing, editors, *New Essays on Belnap–Dunn Logic*, number 418 in Synthese Library, pages 67–86. Springer International Publishing, Switzerland, 2019.

[32] J. Michael Dunn and Nuel D. Belnap, Jr. Homomorphisms of intensionally complemented distributive lattices, (abstract). *Journal of Symbolic Logic*, 32(3):446, 1967.

[33] J. Michael Dunn and Nuel D. Belnap, Jr. Homomorphisms of intensionally complemented distributive lattices. *Mathematische Annalen*, 176:28–38, 1968.

[34] J. Michael Dunn and Gary M. Hardegree. *Algebraic Methods in Philosophical Logic*, volume 41 of *Oxford Logic Guides*. Oxford University Press, Oxford, UK, 2001.

[35] J. Michael Dunn and Robert K. Meyer. Gentzen's cut and Ackermann's gamma. In J. Norman and R. Sylvan, editors, *Directions in Relevant Logic*, pages 229–240. Kluwer, Dordrecht, 1989.

[36] Steve Giambrone. TW_+ and RW_+ are decidable. *Journal of Philosophical Logic*, 14:235–254, 1985.

[37] Paul Halmos and Steven Givant. *Logic as Algebra*. Number 21 in Dolciani Mathematical Expositions. Mathematical Association of America, 1998.

[38] C. I. Lewis and C. H. Langford. *Symbolic Logic*. The Century Co., New York, 1932. 2nd ed., Dover Publications, New York, 1959.

[39] Ricardo Maronna. A characterisation of Morgan lattices. *Portugalie Mathematica*, 23:169–171, 1964.

[40] Robert K. Meyer and J. Michael Dunn. E, R and γ. *Journal of Symbolic Logic*, 34:460–474, 1969. Reprinted in Anderson, A. R. and N. D. Belnap, *Entailment: The Logic of Relevance and Necessity*, vol. 1, Princeton University Press, Princeton, NJ, 1975, §25.2, pp. 300–314.

[41] Robert K. Meyer, J. Michael Dunn, and Hugues Leblanc. Completeness of relevant quantification theories. *Notre Dame Journal of Formal Logic*, 15(1):97–121, 1974.

[42] Robert K. Meyer and Richard Routley. Algebraic analysis of entailment I. *Logique et Analyse*, 15:407–428, 1972.

[43] Sergei P. Odintsov and Heinrich Wansing. The logic of generalized truth values and the logic of bilattices. *Studia Logica*, 103:91–112, 2015.

[44] Hitoshi Omori and Heinrich Wansing, editors. *New Essays on Belnap–Dunn Logic*, volume 418 of *Synthese Library*. Springer International Publishing, Switzerland, 2019.

[45] Helena Rasiowa and Roman Sikorski. *The Mathematics of Metamathematics*. Polish Scientific Publishers, Warszawa, 2nd, revised edition, 1968.

[46] Marlow Sholander. Postulates for distributive lattices. *Canadian Journal of Mathematics*, 3:28–30, 1951.

[47] Yaroslav Shramko. Truth, falsehood, information and beyond: The American plan generalized. In K. Bimbó, editor, *J. Michael Dunn on Information Based Logic*, volume 8 of *Outstanding Contributions to Logic*, pages 191–212. Springer Nature, Switzerland, 2016.

[48] Yaroslav Shramko, J. Michael Dunn, and T. Takenaka. The trilattice of constructive truth values. *Journal of Logic and Computation*, 11:761–788, 2001.

[49] Alasdair Urquhart. The story of γ. In K. Bimbó, editor, *J. Michael Dunn on Information Based Logic*, volume 8 of *Outstanding Contributions to Logic*, pages 93–105. Springer Nature, Switzerland, 2016.

[50] Stephen Yablo. *Aboutness*. Princeton University Press, Princeton, NJ, 2014.

THE ALGEBRA OF INTENSIONAL LOGICS

By

Jon Michael Dunn

A.B., Oberlin College, 1963

Submitted to the Graduate Faculty in the
Division of the Humanities in partial
fulfillment of the requirements
for the degree of
Doctor of Philosophy

University of Pittsburgh

1966

Acknowledgments

This work was supported in part by National Science Foundation Grant GS-689 (History and Philosophy of Science) and by a Dissertation Fellowship from the Woodrow Wilson National Fellowship Foundation.

I wish to thank first, and foremost, my advisor Professor Nuel D. Belnap, Jr. Every page of this study should be considered as footnoted with expressions of gratitude for his invaluable suggestions and encouragement. I also give special thanks to Professor Alan Ross Anderson.

This study has also benefited from talks which I have had with Professors Asenjo, Bertolini, and Rescher of the University of Pittsburgh, with a former fellow student, Professor Meyer of the University of West Virginia, and with fellow students Bas van Frassen, Louis Goble, and Peter Woodruff. Peter Woodruff was especially helpful, and there were times when I thought that he knew more what I was doing than I did.

I have a special debt of gratitude to my wife Sally for her encouragement, especially for her continually telling me that I am not stupid.

Finally, I have both obvious and unobvious debts to my mother and father, and unobvious debts to Professor Henry Koffler of Purdue University, who, although in the biological sciences, first taught me by example what the logical sciences are all about. To them, I dedicate this study.

TABLE OF CONTENTS

iv

I. Introduction

1. Preliminaries on Intensional Logics

It is well-known that both the classical logic of "material implication" and the Lewis logics of "strict implication" contain certain "paradoxes of implication." Among the most notorious of these are that a logical falsehood, e.g., a contradiction, implies any statement whatsoever, and dually, that a logical truth, e.g., an excluded middle, is implied by any statement whatsoever. Such "paradoxes of implication" strike us as paradoxical because it would seem that for one statement to logically imply another, the two statements would have to be relevant to one another.[1] In, recent years, several logical systems have been developed that avoid these paradoxes of relevance. Following a suggestion by Nuel D. Belnap, Jr., we call these systems "intensional logics" because they may be viewed as attempts to explicate a species of implication that in some sense relates the *meanings* of antecedent and consequent.

Church 1951 introduced a purely implicational system, called the "Weak Implicational Calculus," as a relevance preserving system. Ackermann 1956 introduced his $^1\|_2$ system Π′ of "Rigorous (*Strenge*) Implication," which contained machinery for truth functions (negation, disjunction, and conjunction) as well. Anderson and Belnap, in a series of papers commencing with Belnap 1960, developed their system E of "Entailment," which involves a subtle but elegant modification of Ackermann's Π′ so as to obtain what amounts to a deduction theorem. A number of workers have suggested supplementing Church's Weak Propositional Calculus with machinery for truth functions. Thus Prawitz 1965 contains suggestions for such a system of "relevant implication," and Belnap 1965 contains an axiomatization for a system R of "Relevant Implication."[2]

Since it is the systems E and R that will be of primary interest in the sequel, let us without further ado set down axiomatic formulations of these systems. We shall suppose that both E and R are formulated with the logical connectives of implication ($\rightarrow$), disjunction ($\vee$), and negation ($^-$) as primitive, with formation rules as usual. We shall assume that conjunction (&) is defined in the usual De Morgan manner ($A \mathbin{\&} B$ defined as $\overline{\overline{A} \vee \overline{B}}$). The system E may then be defined by the following axiom schemata and

---

[1] For a spirited and insightful defense of this view, cf. Anderson and Belnap 1962b.

[2] Anderson and Belnap credit John R. Bacon for suggesting the name "Relevant Implication" for this system in 1962.

rules.[3] [2]$\|_3$

A.1 $A \rightarrow A$ (Identity)

A.2 $A \rightarrow A \rightarrow B \rightarrow B$

A.3 $A \rightarrow B \rightarrow .B \rightarrow C \rightarrow .A \rightarrow C$ (Transitivity)

A.4 $(A \rightarrow .A \rightarrow B) \rightarrow .A \rightarrow B$ (Contraction)

A.5 $(A \rightarrow \overline{A}) \rightarrow \overline{A}$ (Reductio)

A.6 $A \rightarrow \overline{B} \rightarrow .B \rightarrow \overline{A}$ (Contraposition)

A.7 $\overline{\overline{A}} \rightarrow A$ (Double negation)

A.8 $A \,\&\, B \rightarrow A$ (Conjunction elimination)

A.9 $A \,\&\, B \rightarrow B$ (Conjunction elimination)

A.10 $(A \rightarrow B) \,\&\, (A \rightarrow C) \rightarrow .A \rightarrow B \,\&\, C$ (Conjunction introduction)

A.11 $A \rightarrow A \vee B$ (Disjunction introduction)

A.12 $B \rightarrow A \vee B$ (Disjunction introduction)

A.13 $(A \rightarrow C) \,\&\, (B \rightarrow C) \rightarrow .A \vee B \rightarrow C$ (Disjunction elimination)

A.14 $A \,\&\, (B \vee C) \rightarrow (A \,\&\, B) \vee C$ (Distribution)

A.15 $(A \rightarrow A \rightarrow A) \,\&\, (B \rightarrow B \rightarrow B) \rightarrow .A \,\&\, B \rightarrow A \,\&\, B \rightarrow A \,\&\, B$

R.1 From A and $A \rightarrow B$ to infer B (*Modus ponens*)

R.2 From A and B to infer $A \,\&\, B$ (Adjunction)

This is essentially the original formulation of Belnap 1960, p. 82. We remark that we are not here concerned with independence, and all our formulations will be somewhat redundant for the sake of convenience.

The system R may be obtained from the formulation of E by adding

A.16 $A \rightarrow .A \rightarrow A \rightarrow A$.

This gives essentially the original formulation of Belnap 1965. An interesting alternative formulation of R may be obtained by replacing A.2 with [3]$\|_4$

---

[3]We omit parentheses in favor of dots according to the conventions of Church 1956.

A.2′ $(A \to .B \to C) \to .B \to .A \to C$ (Permutation),

which makes A.15 redundant. Axioms A.1, A.2′, A.3, and A.4, together with *modus ponens* provide the original formulation of Church's Weak Theory of Implication, and they may be easily shown equivalent to axioms A.1–A.4 with *modus ponens*.

A similar alternative formulation of E can be obtained by replacing A.2 with

A.2* $(A \to .B \to C \to D) \to .B \to C \to A \to D$ (Restricted permutation)

Axiom A.2* is just a special case of A.2′, and we shall find this difference between E and R as regards permutation of special interest in Chapter X, when we study the algebraic structure of the two systems.

Belnap 1960, p. 26, showed that the system E is free of paradoxes of relevance by showing that no formula of the form $A \to B$ is a theorem unless A and B have a propositional variable in common. This sharing of a propositional variable represents a connection of meaning. The same proof is easily applicable to R.

The systems E and R are thus quite similar in their treatment of the paradoxes of relevance. They differ primarily in their treatment of what Anderson and Belnap call the "paradoxes of necessity." Anderson and Belnap 1959 observed that a modal structure may be introduced upon Ackermann's system Π' by defining LA (A is necessary) as $A \to A \to A$, and that the structure of the resulting $^4\|_5$ modalities is like that of Lewis's S4. The same is true for E, and Belnap 1960, utilizing an argument of Ackermann 1956, demonstrated that in E no formula of the form $A \to .B \to C$ is a theorem when A is a propositional variable. Anderson and Belnap regard this as a virtue of E because they regard as a "paradox of necessity" a purely contingent proposition's entailing that a proposition is necessary, and where A is a propositional variable it could be interpreted as expressing such a contingent proposition.[4] The system R on the other hand pays no attention to such modal niceties, since A.16 may be read as $A \to LA$, which collapses modalities.

We remark in closing this rather cursory discussion of E and R that the following replacement theorem, which will be of primary importance in elucidating the algebraic structure of these two systems, may be easily proven for both systems by the same argument that Ackermann 1956 used for his system Π'.[5]

---

[4]For extended discussions of the "paradoxes of necessity," cf. Belnap 1960, pp. 6–8, and Anderson and Belnap 1962b, pp. 42–46.

[5]This theorem was first stated for E in Belnap 1960.

Replacement Theorem: Let us say that a formula A is equivalent to a formula B if both $A \rightarrow B$ and $B \rightarrow A$ are theorems. Then if C_A is a formula containing a specific $^5\|_6$ occurrence of the formula A and C_B is the result of replacing this occurrence A by B, and A is equivalent to B, then C_A is equivalent to C_B.

2. Preliminaries on Universal Algebra

In this section we shall rehearse the universal algebraic notions that underlie this study. The purpose of this rehearsal is more to set down a uniform terminology than to provide a grounding in algebra for the uninitiated. This section is based primarily upon Birkhoff's 1948 "Foreword on Algebra," in consultation with Birkhoff 1935 and Rasiowa and Sikorski 1963, pp. 21–31.

Since this dissertation is concerned with the algebra of intensional logics, we had better first make explicit what we understand by an algebra. We define an (*abstract*) *algebra* as a (non-empty) set A of elements, together with a number of operations o_x. Each *operation* o_x is, for some finite number n, a (single-valued) function[6] defined over all n-tuples $(a_1, \dots, a_n)$ of elements of A, which takes values in A, i.e., if o_x is an n-ary operation, then $o_x(a_1, \dots, a_n) \in A$.

By a *subalgebra* of an algebra, we mean a (non-empty) subset B of the algebra that is closed under the operations, $^6\|_7$ i.e., such that if $(b_1, \dots, b_n)$ is an n-tuple of elements of B, and o_x is an n-ary operation, then $o_x(b_1, \dots, b_n) \in B$. The intersection of subalgebras of A is also a subalgebra of A, so for any subset A' of A, there is a least subalgebra B including A'. We call B *the subalgebra generated by* A', and we call A' a *set of generators* of B.

We say of two algebras A and B that they are *similar* if for each n, there is a one-to-one correspondence between their respective n-ary operations. Corresponding operations in similar algebras will customarily be denoted by the same symbol. We may then define a *homomorphism* of one algebra A into a similar algebra B as a mapping of A into B that preserves the corresponding operations. Thus if h is the homomorphism, $h(o_x(a_1, \dots, a_n)) = o'_x(h(a_1), \dots, h(a_n))$. We call a homomorphism that is one-to-one an *isomorphism*.

By an *equivalence relation* on an algebra we mean a binary relation $\equiv$ satisfying each of

1) $a \equiv a$ (Reflexive);

2) $a \equiv b$ implies $b \equiv a$ (Symmetric);

---

[6]We regard all functions (or *mappings*, as we often call them) as single-valued, but sometimes we use this redundancy for emphasis.

3) $a \equiv b$ and $b \equiv c$ imply $a \equiv c$ (Transitive).

By a *congruence relation* on an algebra we mean an equivalence relation $\equiv$ with the Replacement Property for each o_x: If o_x is an n-ary operation, and $a_i \equiv b_i$ for $i < n$, then $o_x(a_1, \ldots, a_n) \equiv o_x(b_1, \ldots, b_n)$.

We recall that every homomorphism h of an algebra A onto an algebra B determines a congruence on A, defining $a \equiv b$ iff $h(a) = h(b)$. We recall likewise that every congruence $^7\|_8$ $\equiv$ on A determines a homomorphism of A onto the *quotient algebra* of A modulo $\equiv$, defined as follows: Let $|a|$ be the set of all elements that are congruent to a under $\equiv$. Let $A/\equiv$ be the family of all such congruence classes, and define operations on $A/\equiv$ so that $o_x(|a_1|, \ldots, |a_n|) = |o_x(a_1, \ldots, a_n)|$. The resulting algebra is obviously an algebra similar to A, and the mapping $h(a) = |a|$ is a homomorphism (called the *natural homomorphism* determined by $\equiv$).

An algebra is said to be *simple* if it has only the two *trivial* congruences, namely the identity relation and the universal relation.

A *direct product* of an indexed set of similar algebras $\{A_y\}_{y \in Y}$ is that algebra whose elements are the indexed sets $\{a_y\}_{y \in Y}$, where each $a_y \in A_y$, and whose operations are defined component-wise, i.e., where $o_x(\{a_{1_y}\}_{y \in Y}, \ldots, \{a_{n_y}\}_{y \in Y}) = \{o_x(a_{1_y}, \ldots, a_{n_y})\}_{y \in Y}$. Direct products are often called *direct unions* by other writers. Obviously the direct product of similar algebras is yet another similar algebra.

For any class K of similar algebras, and for any cardinal n, we may define the *free K-algebra with n free generators*, $FK(n)$, as an algebra in the class K such that any mapping of the a_i into an arbitrary algebra of class K can be extended to a homomorphism of $FK(n)$ into the arbitrary algebra. Trivially $FK(n)$ is unique up to isomorphism. $^8\|_9$

We shall find that when we come to study particular kinds of algebras we shall for the sake of clarity usually define special instances of these general notions, instead of simply letting the general notions be applied directly. The main point of this short excursion into *universal algebra* is that our special definitions for particular algebraic systems are not so special. We now turn our attention to a particular kind of algebra, the lattice, which we shall find to be particularly pervasive in the algebra of logic.

3. Preliminaries on Lattice Theory

This section is based upon Garrett Birkhoff's 1948 *Lattice Theory*, which is deservedly the standard work on the subject. Much as in the proceeding section, we intend here merely a rehearsal of some general algebraic notions that will serve as a framework for our study. Again our aim is a uniform terminology.

We begin by recalling that a partial order on a set is a relation $\leq$ on the set satisfying the following:

1) $a \leq a$ (Reflexive);

2) $a \leq b$ and $b \leq a$ imply $a = b$ (Antisymmetric);

3) $a \leq b$ and $b \leq c$ imply $a \leq c$ (Transitivity).

We may read $a \leq b$ as "a is less than or equal to b," or "b contains a." [9]||[10]

Let us further recall that an *upper bound* of a subset B of a partially ordered set A is an element u of A such that for every $b \in B$, $b \leq u$. The *least upper bound* (*l.u.b.*) of the set B is the upper bound u which is least among upper bounds, i.e., for all upper bounds u' of B, $u \leq u'$. *Lower bounds* and *greatest lower bounds* (*g.l.b.*'s) are defined dually. The l.u.b. of a set consisting of just two elements, a and b, is called a *join* and is denoted by $a \vee b$. The g.l.b. of such a two element set is called a *meet* and is denoted by $a \wedge b$.

Clearly joins and meets need not always exist. But we define a *lattice* as a (non-empty) partially ordered set such that any two of its elements have a join and a meet. A trivial example of a lattice is provided by a set which consists of just a single element a such that $a \leq a$. This is called the *degenerate lattice* **1**, and unless we specifically indicate otherwise, we exclude it when we talk of lattices.

We may define a *complete* lattice as a (non-empty) partially ordered set such that *every* subset (no matter what its cardinality) has a l.u.b. and a g.l.b. We call the l.u.b. of a set B a *generalized join* (sometimes just *join*) and denote it by $\bigvee B$. The g.l.b. is called a *(generalized) meet* and is denoted by $\bigwedge B$. We denote the l.u.b. of an indexed set $\{a_x\}_{x \in X}$ by $\bigvee_{x \in X} a_x$ and its g.l.b. by $\bigwedge_{x \in X} a_x$. Since it is easy to see by induction that every *finite* subset of a lattice has both a l.u.b. [10]||[11] and a g.l.b., trivially all *finite* lattices are at the same time complete.

Although we have defined a lattice as a relational system, it may also be looked upon as an operational system (or *algebra* in the sense of the last section), the operations being meet and join. Indeed, in a lattice it is easy to see that $a \leq b$ iff $a \wedge b = a$ (or $a \vee b = b$). This allows for the following purely operational definition of a lattice, which may be shown equivalent to our relational definition upon defining $a \leq b$ as suggested above. A lattice is a (non-empty) set with two binary operations $\wedge$ and $\vee$, satisfying

L1. $a \wedge a = a$ and $a \vee a = a$ (Idempotent);

L2. $a \wedge b = b \wedge a$ and $a \vee b = b \vee a$ (Commutative);

L3. $a \wedge (b \wedge c) = (a \wedge b) \wedge c$ and $a \vee (b \vee c) = (a \vee b) \vee c$ (Associative);

L4. $a \wedge (a \vee b) = a$ and $a \vee (a \wedge b) = a$ (Absorptive).

It is worth remarking that although a lattice can be looked upon as an algebra, strictly speaking a *complete* lattice generally cannot be so regarded, the reason being that generalized meets and joins may operate upon *infinite* sets, whereas all the operations of an *algebra* are usually regarded to be finitary by definition. However, nothing is lost for our purposes if we regard complete lattices as *generalized algebras* (allowing infinite operations). Then appropriate notions of *generalized subalgebra, generalized homomorphism,* etc., may be defined in obvious analogy to the corresponding definitions for algebras in the last section. $^{11}\|_{12}$

A *sublattice* of a lattice is a (non-empty) subset that is closed under the operations of meet and join. A *complete sublattice* of a complete lattice is analogously a (non-empty) subset that is closed under the operations of generalized meet and generalized join.

Two special types of sublattice are of great importance. The first of these, an *ideal,* is a (nonempty) subset I such that

I1. if $a, b \in I$, then $a \vee b \in I$;   and

I2. if $a \in I$, then $a \wedge b \in I$.

Alternative, and equivalent definitions, may be gotten by either replacing I2 with

I2′. if $a \in I$ and $b \le a$, then $b \in I$,

or by dropping I2 altogether and strengthening I1 to

I1′. $a, b \in I$ *iff* $a \vee b \in I$.

The second of these two special kinds of sublattice, a *filter,* is defined dually.[7] A *filter* is accordingly a (non-empty) subset F such that

F1. if $a, b \in F$, then $a \wedge b \in F$;   and

F2. if $a \in F$, then $a \vee b \in F$.

Equivalently, F2 may be replaced with

F2′. if $a \in F$ and $a \le b$, then $b \in F$,

or F2 may be dropped altogether in favor of the following strengthening of F1: $^{12}\|_{13}$

---

[7] Indeed, Birkhoff 1948, p. 21, uses the term "dual ideal" for this concept.

F1′ $a, b \in F$ *iff* $a \wedge b \in F$.

We shall in the sequel exploit these equivalent definitions of *filter* and *ideal* without special mention.

A trivial example of an ideal (or filter) of a lattice A is A itself. Unless we specifically indicate otherwise, when we speak of ideals (or filters) we shall be meaning to exclude *trivial* ideals (or filters).

An ideal (or filter) is said to be *maximal* if no (non-trivial) ideal (or filter) other than itself includes it.

The definition of *complete ideal* may be obtained by replacing I1 with

CI1. If $A \subseteq I$, then $\bigvee A \in I$,

and the definition of *complete filter* may be obtained by replacing F1 by

CF1. If $A \subseteq F$, then $\bigwedge A \in F$.

A *prime ideal* is an ideal I satisfying

PI. If $a \wedge b \in I$, then $a \in I$ or $b \in I$,

and a *prime filter* is a filter F satisfying

PF. If $a \vee b \in F$, then $a \in F$ or $b \in F$.

A *completely prime ideal* is then an ideal I satisfying

CPI. If $\bigwedge A \in I$, then for some $a \in A$, $a \in I$,

and a *completely prime filter* is then a filter F satisfying

CPF. If $\bigvee A \in F$, then for some $a \in A$, $a \in F$.

Since the intersection of ideals (or filters) is always an ideal (or filter), we may define *the ideal* (or *filter*) *generated by a set* A as the least ideal (or filter) [13][14] including A, and be sure that such exists. When A consists of just a single element a, we speak of the *principal ideal* (or *filter*) *generated by* a, which is the set of all x such that $x \leq a$ ($x \geq a$).

The set-theoretical complement of a prime ideal is a prime filter, and vice versa. The set-theoretical complement of a complete and completely prime ideal is a complete and completely prime filter, and vice versa.

A (*lattice*) *homomorphism* of one lattice $(A, \wedge, \vee)$ into another lattice $(B, \wedge, \vee)$ is a mapping h of A into B such that $h(a \wedge b) = h(a) \wedge h(b)$, and $h(a \vee b) = h(a) \vee h(b)$. A (*lattice*) *isomorphism* is a one-to-one homomorphism.

Stone 1937 observed that prime filters determine in a natural fashion homomorphisms into the two element lattice **2**, whose diagram is as follows:[8]

$$\begin{array}{c} \circ \quad 1 \\ | \\ \circ \quad 0 \end{array}$$

The homomorphism h is determined by a prime filter P so that $h(a) = 1$ if $a \in P$, and $h(a) = 0$ if $a \notin P$. This observation leads to remarkable facts about a certain type of lattice, a *distributive lattice*, which may be defined as a lattice satisfying the following two postulates, [14] $\|_{15}$

DL1. $a \wedge (b \vee c) = (a \wedge b) \vee (a \wedge c)$;   and

DL2. $a \vee (b \wedge c) = (a \vee b) \wedge (a \vee c)$.

Actually, either one of DL1 and DL2 imply the other, and they are both implied by

DL3. $a \wedge (b \vee c) \leq (a \wedge b) \vee c$.

The classic example of a distributive lattice is a *ring of sets*, i.e., a collection of sets that is closed under binary intersections (meets) and binary unions (joins).

Stone 1937 proved, using the axiom of choice, that for elements a and b of a distributive lattice, if $a \not\leq b$, then there exists a prime filter P such that $a \in P$ and $b \notin P$. We shall refer to this result as Stone's Prime Filter Theorem. From this result it may be proven that every distributive lattice is isomorphic to a ring of sets, under the mapping which sends every element into the set of prime filters which contain it (or equivalently into the set of **2** valued homomorphisms which carry it into 1). Since this result itself was first proven in Birkhoff 1933, we shall refer to it as Birkhoff's Representation Theorem for Distributive Lattices. An equivalent statement of this result is that every distributive lattice is isomorphically embeddable in a direct product of **2**.

Similar, and in some ways more striking results, were obtained in Stone 1936 for Boolean algebras. A *Boolean algebra* may be defined as a *complemented* distributive lattice, where a lattice is said to be *complemented* if it has a least element 0 and a greatest element 1, and it [15] $\|_{16}$ satisfies

    C. For any element a, there is an element $\overline{a}$ (the complement of a) such that $a \wedge \overline{a} = 0$, and $a \vee \overline{a} = 1$.

---

[8]For instructions in reading Hasse diagrams, cf. Birkhoff 1948, pp. 5–6.

The classic example of a Boolean algebra is a *field of sets*, i.e., a ring of sets that is closed under set theoretical complementation.

A (*Boolean*) *homomorphism* of one Boolean algebra into another is a lattice homomorphism that preserves complementation as well, i.e., $h(\bar{a}) = \overline{h(a)}$, and a (*Boolean*) *isomorphism* is a one-to-one homomorphism.

Stone 1936 proved, again using the axiom of choice, that for any two elements a and b of a Boolean algebra, if $a \not\leq b$, then there exists a maximal filter M such that $a \in M$ and $b \notin M$ (actually, this is just a special case of his later 1937 result for distributive lattices, which we have mentioned earlier, since in a Boolean algebra prime filters and maximal filters coincide). Stone further observed that a maximal filter M of a Boolean algebra has the property that for every element a, M contains exactly one of a and $\bar{a}$, and that hence maximal filters determine in a natural fashion homomorphisms into the Boolean algebra **2**. Stone used these facts to show that every Boolean algebra is isomorphic to a field of sets, and equivalently that every Boolean algebra is isomorphically embeddable in a direct product of **2**.

Before we close this brief summary of lattice theoretical notions that are relevant to logic, we shall $^{16}\|_{17}$ briefly mention two other types of lattices. The first of these is a *closure algebra*, which is a Boolean algebra together with a unary operation C that satisfies the following Kuratowski closure postulates:

K1. $a \leq Ca$;

K2. $CCa = a$;

K3. $C(a \vee b) = Ca \vee Cb$;   and

K4. $C(0) = 0$.

The notion of a closure algebra is a generalization of the notion of a topological space; and indeed if the underlying Boolean algebra is restricted to be a field of *all* subsets of some set, then the resulting notion coincides exactly with Kuratowski's definition of a topological space. McKinsey and Tarski 1944 proved that every closure algebra is isomorphically embeddable in a topological space, which result corresponds to Birkhoff's and Stone's representations for distributive lattices and Boolean algebras.

We now define the other type of lattice. A lattice is said to be *relatively pseudo-complemented* if for every pair of elements a, b there exists an element $a \Rightarrow b$ such that for all elements x, $x \leq a \Rightarrow b$ iff $x \wedge a \leq b$. Every relatively pseudo-complemented lattice may be shown to be distributive and to have a greatest element 1 (indeed, for all a, $a \Rightarrow a = 1$). A relative pseduo-complemented lattice that has a least element 0 as well is called a pseudo-Boolean algebra.[9] [17]‖[18]

The element $a \Rightarrow 0$ is called the pseudo-complement of a, and is denoted by $-a$.

The set of all open sets in a topological space serves as an illustrative example of a pseudo-Boolean algebra, where meet is intersection, join is union, and where $I(a)$ (the *interior* of a) is defined as $\overline{C(\overline{a})}$, $a \Rightarrow b = I(\overline{a} \vee b)$, and $-a = I(\overline{a})$. Indeed, McKinsey and Tarski 1946 showed in effect that every pseudo-Boolean algebra is isomorphically embeddable in the pseudo-Boolean algebra of all open sets of some topological space. Pseudo-Boolean algebras are thus intimately connected with closure algebras, since every pseudo-Boolean algebra can be embedded in a closure algebra.

4. Algebraic Logic

In this section we shall recall some of the high points in the development of algebraic logic, our aim being to provide a framework of established results concerning non-intensional logics with which our subsequent treatment of the algebra of intensional logics may be compared. Although we shall chiefly be discussing the algebra of the classical propositional calculus, this discussion is intended to have a certain generality. We mean to emphasize the essential features of the relation of the classical propositional calculus to Boolean algebra, remarking from time to time what is special to this relation and what is [18]‖[19] generalizable to the algebra of other propositional calculuses. It should be mentioned that we here restrict ourselves to the algebra of propositional logics, despite the fact that profound results concerning the algebra of the classical predicate calculus have been obtained by Tarski, Halmos, and others. We do this because most of our results concerning intensional logics deal with their propositional formulations. It should also be mentioned that we are not here concerned with setting down the history of algebraic logic, and that, much as in a historical novel, historical figures will be brought in mainly for the sake of dramatic emphasis. The interested reader may refer to Rasiowa and Sikorski 1963 for proofs and proper citations of most of the results we discuss.

About the middle of the last century, the two fields of abstract algebra and symbolic logic came into being. Although algebra and logic had been around for some time, *abstract* algebra and *symbolic* logic were essentially new developments. Both these fields owe their origins to the insight that formal systems may be investigated without explicit recourse to their intended interpretations.

---

[9]McKinsey and Tarski 1946 studied the dual lattice, which they called a *Brouwerian algebra*. By analogy, pseudo-Boolean algebras are often called *Heyting algebras*.

This insight led George Boole, in 1847 with his *Mathematical Analysis of Logic*, to formulate at one and the same time perhaps the first example of a non-numerical algebra and the first example of a symbolic logic. He observed that the operation of conjoining two $^{19}\|_{20}$ propositions[10] had certain affinities with the operation of multiplying two numbers. He saw that by letting letters like '*a*' and '*b*' stand for propositions, just as they stand for numbers in ordinary algebra, and that by letting juxtaposition of letters stand for the operation of conjunction, just as it stands for multiplication in ordinary algebra, these affinities could be brought to the fore. Thus, for example, $ab = ba$ is a law of this *algebra of logic*, just as it is a law of the ordinary algebra of numbers. At the same time, the algebra of logic has certain differences from the algebra of numbers since, for example, $aa = a$. The differences are just as important as the similarities, for whereas the similarities suggested a truly *symbolic* logic, like the "symbolic arithmetic" that comprises ordinary algebra, the differences suggested that algebraic methods could be extended far beyond the ordinary algebra of numbers.

Oddly enough, despite the fact that Boole's algebra was thus connected with the origins of both abstract algebra and symbolic logic, the two fields developed for some time thereafter in comparative isolation from one another. On the one hand, the notion of a Boolean algebra was perfected $^{20}\|_{21}$ by Jevons, Schröder, Huntington, and others (until it reached the modern conception given in the last section), and developed as a part of the growing field of abstract algebra. On the other hand, the notion of a symbolic logic was developed along subtly different lines from Boole's original algebraic formulation, starting with Frege and receiving its classic statement in Whitehead and Russell's *Principia Mathematica* of 1910. The divergence of the two fields was partly a matter of attitude. Thus Boole, following in the tradition of Leibnitz, wanted to study the mathematics of logic, whereas the aim of Frege, Whitehead, and Russell was to study the logic of mathematics. The modern field of mathematical logic, of course, recognizes both approaches as methodologically legitimate, and indeed embraces them both under the very ambiguity of its name, "mathematical logic," but the Frege–Whitehead–Russell aim to reduce mathematics to logic obscured for some time the two-headedness of the mathematical-logical coin.

There is more that a difference in attitude, however, between Boole's algebraic approach to logic, and the Frege–Whitehead–Russell approach to logic, which for want of a better word we shall call logistic. We shall attempt

---

[10]Boole tended to identify propositions with classes, the conjunction of propositions thus corresponding to the intersection of classes. Boole's algebra of logic is thus at the same time an algebra of classes, but we shall ignore this aspect of Boole's algebra in the present discussion.

to bring out this difference between the two approaches, which was either so profound, or so subtle, that the precise connection between the two ways of looking at logic was not $^{21}\|_{22}$ discovered until the middle 1930's.[11]

Let us begin by looking at a logistic presentation of the classical propositional calculus that is essentially the same as in *Principia Mathematica* except that we use axiom schemata and thereby do without the rule of substitution, which was tacitly presupposed by *Principia*. This presentation begins by assuming that we have a certain stock of propositional variables p, q, r, etc., and then specifies that these are (well-formed) formulas and that further formulas may be constructed from them by the usual inductive insertion of logical connectives (and parentheses). The particular logical connectives assumed in this presentation are those of disjunction ($\vee$) and negation ($^{-}$), although conjunction is assumed to be defined in terms of these so that $A \,\&\, B$ is an abbreviation for $\overline{\overline{A} \vee \overline{B}}$, and material implication is also assumed to be defined so that $A \supset B$ is an abbreviation for $\overline{A} \vee B$. A certain proper subset of these formulas are then singled out as axioms. These axioms are all instances of the following axiom schemata:

1. $(A \vee A) \supset A$

2. $B \supset (A \vee B)$

3. $(A \vee B) \supset (B \vee A)$ $^{22}\|_{23}$

4. $(A \vee (B \vee C)) \supset (B \vee (A \vee C))$

5. $(B \supset C) \supset ((A \vee B) \supset (A \vee C))$

These axioms are called theorems, and it is further specified that additional formulas are theorems in virtue of the following rule:

Modus ponens. If A is a theorem, and if $A \supset B$ is a theorem, then B is a theorem.

The point of this perhaps too tedious but not too careful rehearsal of elementary logic is to give us some common ground for a comparison of the classical propositional calculus with a Boolean algebra. There are certain surface similarities that are misleading. Thus, for example, a Boolean algebra has certain items called elements which are combined by certain operations to give other elements, just as the classical propositional calculus has certain items called formulas which are combined by the operation of

---

[11] The difference we have in mind is essentially the distinction that Curry 1963 pp. 166–168, makes between a *relational* (algebraic) system and an *assertional* (logistic) system, though we shall have to be more informal than Curry since we do not have his nice formalist distinctions at hand.

inserting logical connectives to give other formulas. They are both then, from this point of view, abstract algebras in the sense of section 2. This fact might lead one to confuse the operation of disjoining two formulas A and B so as to obtain $A \vee B$, with the operation of joining two elements of a Boolean algebra a and b so as to obtain $a \vee b$. There are essential differences between these two binary operations. Consider, for example, that where A is a formula, $A \vee A$ is yet another *distinct* formula since $A \vee A$ contains at least one more occurrence of the disjunction sign $\vee$ than does A. Yet in a Boolean $^{23}\|_{24}$ algebra, where a is an element, $a \vee a = a$, by L1. Further, in the algebra of formulas, where A and B are distinct formulas, the formula $A \vee B$ is distinct from the formula $B \vee A$ since although the two formulas are composed of the same signs, the signs occur in different orders. Yet in a Boolean algebra, $a \vee b = b \vee a$, by L2.

The trouble with the algebra of formulas is that like the bore at a party, it makes too many distinctions to be interesting. Its detailed study might be of interest to the casual thrill-seeker who is satisfied with "something new everytime," but the practiced seeker of identity in difference demands something more than mere newness. To such a seeker as Boole, the "identity" of two such different formulas as $A \vee A$ and A, or $A \vee B$ and $B \vee A$, lies in the fact that they express the "same proposition," but this was only understood at such an intuitive level until the 1930's, when Lindenbaum and Tarski made their explication of this insight.

Lindenbaum and Tarski observed that the logistic, presentation of the classical propositional calculus could be made to reveal a deeper algebra than the algebra of formulas that it wore on its sleeve. Their trick was to introduce a relation of *logical equivalence* $\equiv$ upon the class of formulas by defining $A \equiv B$ iff both $A \supset B$ and $B \supset A$ are theorems. It is easy to show that the relation $\equiv$ is a genuine equivalence relation. Thus reflexivity follows because $A \supset A$ is a theorem, symmetry follows by definition, $^{24}\|_{25}$ and transitivity follows from the fact that whenever $A \supset B$ and $B \supset C$ are theorems, then $A \supset C$ is a theorem (the rule form of transitivity). It is interesting to observe that since the classical propositional calculus has a "well-behaved" conjunction connective, i.e., $A \,\&\, B$ is a theorem iff both A and B are theorems, then the same effect may be gotten by defining $A \equiv B$ iff $(A \supset B) \,\&\, (B \supset A)$ is a theorem. It is natural to think of the class of all formulas logically equivalent to A, which we represent by $|A|$, as one of Boole's "propositions." Operations are then defined upon these equivalence classes, one corresponding to each logical connective, so that $\overline{|A|} = |\overline{A}|$, $|A| \vee |B| = |A \vee B|$, $|A| \,\&\, |B| = |A \,\&\, B|$, and $|A| \supset |B| = |A \supset B|$.[12] Since the

---

[12]Observe that in the classical propositional calculus, the last two operations may actually be defined in terms of the first two since conjunction and material implication

Replacement Theorem holds for the classical propositional calculus,[13] these operations may be shown to be genuine (single-valued) operations. The point of the Replacement Theorem is to ensure that the result of operating upon equivalence classes does not depend upon our choice of representatives for the classes. Thus, for example, if $A \equiv B$, then $|A| = |B|$. But then for the unary operation corresponding to negation to be single-valued, we must have $\overline{|A|} = \overline{|B|}$, i.e., $^{25}\|_{26}$ $|\overline{A}| = |\overline{B}|$, i.e., $A \equiv B$, which is just what the Replacement Theorem guarantees us.

Let us call the algebra so defined the *Lindenbaum algebra* of the classical propositional calculus.[14] It is simply a matter of axiom-chopping to see that this is a Boolean algebra. Thus, for example, it is easy to see that $|A| \vee |A| = |A|$, even though $A \vee A$ and A are distinct formulas, for $(A \vee A) \supset A$ is an instance of axiom schema 1, and $A \supset (A \vee A)$ is an instance of axiom schema 2. Similarly, $|A| \vee |B| = |B| \vee |A|$ follows from two instances of axiom schema 3. The other laws of a Boolean algebra may be established analogously. Let us observe, as might have been expected, that $|A| \leq |B|$ iff $A \supset B$ is a theorem.

The essentials of the Lindenbaum–Tarski method of constructing an algebra out of the classical propositional calculus can be applied to most other well-motivated propositional calculuses, and because of the intuitive properties of conjunction and disjunction, most of the resulting Lindenbaum algebras are lattices, indeed, distributive lattices.[15] In particular, the Lindenbaum algebra of Lewis's $^{26}\|_{27}$ modal logic S4 is a closure algebra, and the Lindenbaum algebra of Heyting's intuitionist logic is a pseudo-Boolean algebra.[16] One of the most remarkable features of the reunion of logic and algebra that took place in the 1930's was this discovery that certain non-classical propositional calculuses that had captured the interest of logicians

---

[13] Cf. the statement of the Replacement Theorem for E in section 1, reading $\to$ as $\supset$.

[14] We follow Rasiowa and Sikorski 1963, p. 245n, in calling this device a *Lindenbaum algebra*, despite the fact that it first appeared in print in Tarski 1935, for essentially the reasons they give.

[15] Various logicians, at various times, have, however, questioned the various principles needed for the construction of a Lindenbaum algebra, and some logicians have even developed logical systems that do not have these principles. For example, Strawson 1952, p. 15, has cast aspersion on the law of identity (though he seems prepared to accept it as a "technical" device). Smiley 1959 has worked out a theory of non-transitive "entailment." Fitch's 1952 system apparently does not have the Replacement Theorem. And there are some systems containing logical connectives for "conjunction" and "disjunction" which are not lattices. Thus, for example, both Angell's 1962 system of the "subjunctive conditional" and McCall's 1965 system of "connexive implication" are constructed so that $A \& B$ does not always imply A. The suggestion of a non-distributive logic of quantum mechanics may be found in Birkhoff and von Neumann 1936.

[16] Cf. McKinsey 1941, McKinsey and Tarski 1948, and Birkhoff 1948, pp. 195–196.

may be defined in terms of disjunction and negation.

had such intimate connections with certain structures that had been developed by algebraists in the context of lattice theory—a generalization of the theory of Boolean algebras that by then stood on its own.

An even more striking example of the identification of notions and results that were of independent interest to both logicians and algebraists may be found in Tarski's 1935 theory of deductive systems, which was later seen to overlap the Boolean ideal theory of Stone 1936. Apparently Tarski did not realize the algebraic significance of his theory until he read Stone, and conversely, Stone did not realize the logical significance of his theory until he read Tarski.[17] [27]||[28]

Intuitively, a *deductive system* is an extension of a logistic presentation of a propositional calculus (assumed not to have a rule of substitution) that has been obtained by adding additional formulas as axioms (however, Tarski explicitly defined the notion only for the classical propositional calculus). Stone defined a (lattice) ideal as we did in the preceding section, and at the same time showed that Boolean algebras could be identified with idempotent rings (with identity) the so-called *Boolean rings*, and that upon this identification the (lattice) ideals were the ordinary ring ideals. This identification was of great importance since the value of ideals in ring theory was already well-established, the concept of an ideal having first been developed by Dedekind as an explication of Kummer's "ideal number," which arose in connection with certain rings of numbers (the algebraic integers). It is a tribute to the powers of abstract algebra that the abstract concept of an ideal can be shown to underlie both certain number theoretical concepts and certain logical concepts.

The connection between deductive systems and ideals becomes transparent upon the Lindenbaum identification of a formula with its logical equivalents. Then a deductive system is the dual of an ideal, namely, what we call a *filter*, and conversely, a filter is a deductive system. Without going into the details of this connection, let us simply remark the analogy between a deductive system and a [28]||[29] filter. Let us assume that F is a set of theorems of some extension of the classical propositional calculus, or of almost any well-known, well-motivated propositional calculus. Then both formal and intuitive considerations demand that if $A, B \in F$, then $A \& B \in F$, which corresponds to property F1 of our definition of a filter, and that if $A \in F$, then $A \vee B \in F$, which corresponds to our property F2. It is curious to observe that if we consider the set of *refutable* formulas, i.e., those formulas whose negations are theorems, then we get an ideal in the Lindenbaum algebra. The fact that theorems are more customary objects for logical study than refutables, while at the same time ideals are more cus-

---

[17]Cf. Kiss 1961, pp. 5–6.

tomary objects for algebraic study than filters, has led Halmos 1962, p. 22, to conjecture that the logician is the dual of the algebraist. By duality we obtain as a corollary that the algebraist is the dual of the logician.

Upon the Lindenbaum identification of logically equivalent formulas, the filter of theorems of the classical propositional calculus has a particularly simple structure, namely, it is the trivial filter that contains just the 1 of the Boolean algebra that so results. This fact depends upon one of the paradoxes of implication, namely, that where B is a theorem, then $A \supset B$ is a theorem. This means that all theorems are logically equivalent and hence identified with each other in the same equivalence class, and that any theorem is logically implied by any formula and hence this $^{29}\|_{30}$ equivalence class of theorems ends up at the top of the Boolean algebra. In short, A is a theorem iff $|A| = 1$. This explicates a notion of Boole's that a proposition a is a logical truth iff $a = 1$. Since the same paradox of implication is shared with many other propositional calculuses, e.g., S4 and the intuitionist logic, this algebraically elegant characterization of theoremhood is widely applicable. But since in the intensional logics that we shall be studying it is not the case that all theorems are logically equivalent (cf. section 1), we shall have to use a different algebraic analogue of theoremhood. Note that we can always resort to the inelegant characterization that A is a theorem iff $|A|$ is in the Lindenbaum analogue of the deductive system based on the logic. This means, in the case of the intensional logics that we shall be studying, that the algebraic analogue of the class of theorems is the filter generated by the elements that correspond to the axioms, although we shall find a more elegant way of putting this in Chapter X. The same characterization actually holds for the Lindenbaum algebra of the classical propositional calculus, its being but a "lucky accident," so to speak, that this filter is the trivial filter that may hence be thought of as identical with the element 1 that is its sole member. The algebra of intensional logics is thus demonstrably "non-trivial."

So far we have been discussing the "algebra of the *syntactics* of a propositional logic since the notions of $^{30}\|_{31}$ *formula, theorem,* etc., by which the Lindenbaum algebra is defined, all ultimately depend only upon the syntactic structure of sequences of signs of the system. But there is another side to logic, namely, *semantics*, which studies the interpretations of logical systems. Thus, to use a well-known example, to say of the formula $A \vee \overline{A}$ that it is a theorem of the classical propositional calculus is to say something syntactical, whereas to say of $A \vee \overline{A}$ that it is a tautology is to say something semantical since it is to say something about the formula's interpretations in the ordinary two-valued truth tables, namely, that its value is *true* under every valuation. Now we have already discussed an algebraic way of expressing the first fact, namely, we can say that $|A \vee \overline{A}| = 1$. What we now want is an algebraic way of expressing the second fact. It is well-known that the ordinary truth tables may be looked at as the two element Boolean algebra **2** (where *true* is 1 and *false* is 0). This allows us to define a *valuation* into **2** (or any Boolean algebra) as a mapping of the formulas into the Boolean algebra that carries negation into complementation, disjunction into join, etc., all in the obvious way. We can then define a formula A as *valid* with respect to a class of Boolean algebras iff for every valuation into a Boolean algebra in the class, $v(A) = 1$. We can define the classical propositional calculus as *consistent* with respect to a class of Boolean algebras iff every theorem is valid with respect to that class, and as *complete* with [31]$\|_{32}$ respect to the class iff every formula that is valid with respect to the class is a theorem. Observe that these definitions coincide with the usual definitions with respect to truth tables when the class of Boolean algebras in question consists of just the single Boolean algebra **2**. Observe also that similar definitions may be given for non-classical propositional calculuses once the appropriate algebraic analogue of theoremhood has been picked out.

It may easily be shown that the classical propositional calculus is both consistent and complete with respect to the class of all Boolean algebras. Thus consistency may be shown in the usual inductive fashion, showing first that the axioms are valid, and then that the rules (*modus ponens*) preserve validity. Completeness is even more trivial, since it may be immediately seen that if a formula A is not a theorem, then if we define for every formula B, $v(B) = |B|$, that under this valuation $v(A) \neq 1$. Of course, this completeness result is not as satisfying as the more familiar two-valued result since, among other things, it does not immediately lead to a decision procedure (the Lindenbaum algebra of the classical propositional calculus formulated with an infinite number of propositional variables not being finite). But it does form the basis for an algebraic proof of the two-valued result. We shall see this after a short digression concerning valuations and homomorphisms.

The notion of a homomorphism is the algebraic analogue of a valuation. From any valuation v of the classical [32]$\|_{33}$ propositional calculus into a Boolean algebra B we can define a homomorphism h of the Lindenbaum algebra into B as $h(|A|) = v(A)$, and conversely, from any homomorphism h of the Lindenbaum algebra we can define a valuation v as $v(A) = h(|A|)$. The second fact is obvious, but the first fact requires a modicum of proof, which is not without intrinsic interest. What needs to be shown is that the function h is well-defined in the sense that its value for a given equivalence class as argument does not depend upon our choice of a formula as representative of that equivalence class, i.e., that if $A \equiv B$, then $v(A) = v(B)$.

This amounts to a special case of the semantic consistency result, for what must be shown is that if $A \supset B$ is a theorem, then $v(A) \leq v(B)$, i.e., $v(A) \supset v(B) = v(A \supset B) = 1$. The fact that every valuation thus determines a homomorphism allows us to observe that the Lindenbaum algebra of the classical propositional calculus formulated with n propositional variables is the free Boolean algebra with n free generators. Note that it is typical of algebraic logic that no artificial restrictions are placed upon the assumed cardinality of the stock of propositional variables. Although there may be very good metaphysical or scientific reasons for thinking that the number of actual or possible physical inscriptions of propositional variables is at most denumerable, still the proof we are about to sketch is not affected by questions of cardinality. [33] $\|_{34}$

The proof begins by observing that distinct propositional variables determine distinct equivalence classes. Let us suppose that the propositional variables are p_x, and that we have a mapping f of their equivalence classes $|p_x|$ into a Boolean algebra B. We can then define a new function s from the propositional variables into B by $s(p_x) = f(|p_x|)$. This function s then inductively determines a valuation v into B, and the valuation v in turn determines a homomorphism h of the Lindenbaum algebra into B, as we have just seen.

The situation we have described above is typical of the algebra of logic. We take a logic and form its Lindenbaum algebra (if possible). We then abstract the Lindenbaum algebra's logical structure and find a class of algebras such that the Lindenbaum algebra is free in the class. That the Lindenbaum algebra is *in* the class then amounts to the logic's completeness, and that it is free in the class amounts to the logic's consistency. The trick is to abstract the Lindenbaum algebra's logical structure in an interesting way. Thus, for example, it is interesting that the Lindenbaum algebra of S4 is free in the class of closure algebras, and it is interesting that the Lindenbaum algebra of the intuitionist logic is free in the class of pseudo-Boolean algebras, because these algebras are rich enough in structure and in applications to be interesting in their own right. Let us remark that it is irrelevant whether the logic or the algebra comes first in the actual [34] $\|_{35}$ historical process of investigation.

Having thus picked an appropriate class of algebras with respect to which the logic may be shown consistent and complete, it is, of course, desirable to obtain a sharper completeness result with respect to some interesting subclass of the algebras. One perennially interesting subclass consists of the finite algebras, for then a completeness result leads to a decision procedure for the logic. McKinsey 1941 and McKinsey and Tarski 1948 have obtained such finite completeness results for S4 with respect to closure algebras, and for the intuitionist logic with respect to pseudo-Boolean algebras.

It might be appropriate to point out that due to the typical coincidence of valuations and homomorphisms, algebraic semantics may be looked at as a kind of algebraic *representation theory*, representation theory being the study of mappings, especially homomorphisms, between algebras. This being the case, one cannot expect to obtain deep completeness results from the mere hookup of a logic with an appropriate class of algebras unless that class of algebras has an already well-developed representation theory. Of course, the mere hookup can be a tremendous stimulus to the development of a representation theory, as we shall find when we begin our study of the algebra of intensional logics.

We close this section with an example of how a well-developed representation theory can lead to deep completeness results. We shall show how certain representation [35]‖[36] results for Boolean algebras of Stone 1936, dualized for the sake of convenience from the way we reported them in the last section to the way Stone actually stated them, lead to an elegant algebraic proof of the completeness of the classical propositional calculus with respect to **2**. Of course, in point of fact the completeness result (with respect to truth tables) was first obtained by Post 1930 by a non-algebraic proof using cumbersome normal form methods, but this is irrelevant to the point being made.

We shall show that a formula A is valid (in **2**) only if it is a theorem by proving the contrapositive. We thus suppose that A is not a theorem, i.e., that $|A| \neq 1$. By a result of Stone's we know that there is a maximal ideal M in the Lindenbaum algebra such that $|A| \in M$. But also by a result of Stone's we know that there is a homomorphism h of the Lindenbaum algebra that carries all members of M into 0. Thus $h(|A|) = 0$. Now through the connection between homomorphisms and valuations, we can define a valuation v into **2** by $v(A) = h(|A|)$, and thus there is a valuation v such that $v(A) = 0$, i.e., $v(A) \neq 1$, which completes the proof.

Even more remarkable connections between Stone's results and the completeness of the classical propositional calculus with respect to **2** have been obtained. Thus, for example, Henkin 1954 has essentially shown that Stone's representation theorem for Boolean algebras is directly equivalent to the completeness theorem stated in slightly [36]‖[37] stronger form than we have stated it (cf. also Łoś 1957 for critical modifications).

Let us remark several distinctive features of the "algebraic" proof of the completeness theorem that we have given that make it algebraic. It uses not only the language of algebra, but also the results of algebra. The only role the axioms and rules of the classical propositional calculus play in the proof is in showing that the Lindenbaum algebra is a Boolean algebra, and hence that the Boolean results may be applied. Further, the proof is wildly transfinite. By this we mean not only that no assumptions have been made regarding the cardinality of the propositional variables, but also that a detailed examination of Stone's proof regarding the existence of the essential maximal ideal would reveal that he used the axiom of choice. The proof is at the same time wildly non-constructive, for we are given no way to construct the crucial valuation. A Lindenbaum algebra is thus treated by the same methods as any algebra. Let us say in closing that although there may be philosophical objections to such methods of proof, these objections cannot be directed at just algebraic logic, but instead must be directed at almost the whole of modern algebra.

5. The Algebra of Intensional Logics

In this section we discuss some of the more important definitions and results that have been previously developed [37]$\|_{38}$ and applied to the algebra of the intensional logics E and R. Most of what we say here will be expanded later as the occasion arises.

In Belnap and Spencer 1964 there was introduced the notion of an *intensionally complemented distributive lattice with truth-filter* (*icdl w/t-f*) as a quadruple $(A, \leq, N, T)$ (where A is a set, $\leq$ a relation on A, N a function on A, and T a subset of A) that satisfies the following:

(DL) A is a distributive lattice under $\leq$;

(N1) $NNa = a$;

(N2) if $a \leq b$, then $Nb \leq Na$;

(T) T is a *truth-filter*, i.e., T is a filter that both consistent in the sense that there is no $a \in A$ such that both $a \in T$ and $Na \in T$, and exhaustive in the sense that for every $a \in A$, either $a \in T$ or $Na \in T$ (observe that this condition implies that T is prime).

The operation N is called *intensional complementation*. Let us remark that it easily follows from (N1) that N is a one-to-one mapping of A onto itself, and that it easily follows from this, together with (N2), that $N(a \wedge b) = Na \vee Nb$ and $N(a \vee b) = Na \wedge Nb$ (the so-called *De Morgan laws*). Such a one-to-one mapping of a lattice onto itself that satisfies the De Morgan laws has been called a dual automorphism in Birkhoff 1948, and when such a mapping also satisfies (N1), i.e., is of *period two*, Birkhoff calls it an involution. It is easy to see that (N1) and (N2) are equivalent to the requirement that N be an involution. [38]$\|_{39}$

The operation of Boolean complementation is an example of an involution. An icdl w/t-f may thus be looked at as a generalization of a (non-degenerate) Boolean algebra, the truth-filter T corresponding to a designated maximal filter of the Boolean algebra (cf. section 3). Icdl's w/t-f differ most markedly from Boolean algebras in that it is not in general the case that $a \wedge \mathrm{N}a \leq b$, nor dually, that $a \leq b \vee \mathrm{N}b$. This makes them ideally suited for algebraic investigations of the intensional logics E and R since, as we have pointed out in section 1, these systems do not contain the corresponding paradoxes of implication, namely, that a contradiction implies anything, and dually, that anything implies an excluded middle.

Belnap 1965 has accordingly formulated appropriate semantic notions in terms of icdl's w/t-f, and shown a certain fragment of E and R appropriately consistent and complete. Belnap 1965 actually algebraizes, and at the same time extends, certain notions and results developed in Anderson and Belnap 1962 and 1963. In these semantic applications of icdl's w/t-f, a certain especially simple icdl w/t-f, which Belnap 1965 calls M_0 plays a fundamental role similar to that played by the Boolean algebra **2** in the semantics of the classical propositional calculus. Thus corresponding to the well-known fact that a formula is a theorem of the classical propositional calculus iff it is valid in **2**, we have the fact that a *first degree entailment* (a formula of the form $A \to B$, where A and B are purely [39]‖[40] truth-functional) is a theorem of E (or R) iff it is valid in M_0. This result was obtained by a Post-style "normal form" proof in Anderson and Belnap 1962a, and by a Kanger-style "tree" proof in Anderson and Belnap 1963. We give an algebraic proof in chapter VI.

Anderson and Belnap's result leads to an immediate decision procedure for the first degree entailment fragment of E and R since the crucial icdl w/t-f M_0 has the following finite diagram, where $N(\pm a) = \mp a$, and $T = \{+0, +1, +2, +3\}$ (the labeling derives from its matrix presentation in Belnap 1960).

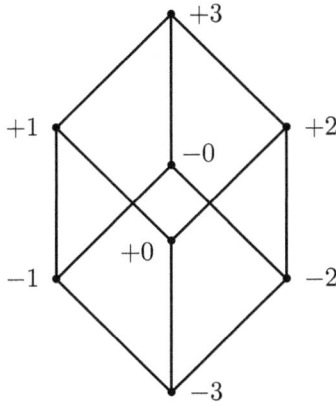

Anderson and Belnap 1963 and Belnap 1965 also obtained deep semantic results for the corresponding fragments of E and R formulated with quantifiers. In this connection, Belnap 1965 gave a complete generalization of an icdl w/t-f as follows. An icdl w/t-f is said to be *complete, completely distributive*, and *with complete truth filter*, provided, respectively, the following conditions hold:

(C) A is a complete lattice;

(CD) $\bigwedge_{x \in X} \bigvee_{y \in Y} a_{x,y} = \bigvee_{f \in U} \bigwedge_{x \in X} a_{x,f(x)}$, and dually, where U is the set of all functions from X into Y.

(CT) T is a complete filter (observe that this condition implies that the truth-filter T is completely prime).

Since in the sequel we do not consider icdl's w/t-f which are complete in the sense of (C), but fail to satisfy (CD) and (CT), we shall permit ourselves to mean by a *complete icdl w/t-f* one satisfying all of (C), (CD), and (CT).

In Belnap and Spencer 1964 there was also introduced the notion of an *intensionally complemented distributive lattice* (*icdl*), that is subtly different than the notion of intensionally complemented distributive lattice *with truth filter* (icdl w/t-f), which was introduced in the same paper and which we have already described. An *icdl* is a triple $(A, \leq, N)$ such that for *some* subset T of A, $(A, \leq, N, T)$ is an icdl w/t-f. Structures like icdl's except that it is not required that there exists a truth-filter, i.e., triples [41]||[42] $(A, \leq, N)$ satisfying (DL), (N1), and (N2), but not necessarily (T), have been called *De Morgan* lattices in Monteiro 1960. Such structures have also been called quasi-Boolean algebras in Białynicki-Birula and Rasiowa 1957[18] and *distributive involution- (i-) lattices* in Kalman 1958, in which papers certain representation and embedding results were obtained which we shall have reason to discuss in chapters III and VI.

In Belnap and Spencer 1964 it was shown that for an arbitrary De Morgan lattice, the condition that it be an icdl is equivalent to the condition that N have no fixed point, i.e., condition (N3). This result includes as a special case the well-known fact due to Stone 1936 that a Boolean algebra has a maximal filter iff it is non-degenerate, since the degenerate Boolean algebra 1 is the only one with an element a such that $a = \bar{a}$.

In Belnap 1965, a *complete icdl* was defined to be a triple $(A, \leq, N)$ such that for *some* subset T of A, $(A, \leq, N, T)$ is a complete icdl w/t-f. In Belnap and Spencer 1964 it was shown that there are De Morgan lattices that satisfy (C) and (N3), and yet still have no truth filter T that satisfies (CT). In Belnap 1965 it was announced that adding (CD) does guarantee the existence of such a T. We $^{42}\|_{43}$ accordingly define *a complete De Morgan lattice* as one satisfying (C) and (CD). In the next chapter we shall give a formulation of Belnap's proof.

Let us define $(A', \leq, N)$ to be a *De Morgan sublattice* of a De Morgan lattice $(A, \leq, N)$ iff $(A', \leq)$ is a sublattice of $(A, \leq)$ and A is closed under N, i.e., if $a \in A$, then $Na \in A$. When $(A, \leq, N)$ is an icdl, we shall call its De Morgan sublattice $(A', \leq, N)$ a *sub-icdl*. Note that a De Morgan sublattice is a De Morgan lattice, and that a sub-icdl is an icdl (this last follows from the fact that if A has a truth-filter T, then A' has a truth filter $T' = A' \cap T$). Let us define $(A', \leq, N, T')$ to be a *sub-icdl w/t-f* of an icdl w/t-f $(A, \leq, N, T')$ iff $(A', \leq, N)$ is a sub-icdl of $(A, \leq, N)$ and $T' = A' \cap T$.

We may then similarly define $(A', \leq, N)$ to be a *complete De Morgan sublattice* of a complete De Morgan lattice $(A, \leq, N)$ iff $(A', \leq)$ is a complete sublattice of $(A, \leq)$ that is closed under N. When $(A, \leq, N)$ is a complete icdl, we shall call its complete De Morgan sublattice $(A', \leq, N)$ a *complete sub-icdl*. We note that a complete De Morgan sublattice is a complete De Morgan lattice, and that a complete sub-icdl is a complete icdl. Let us define $(A', \leq, N, T)$ to be a *complete sub-icdl w/t-f* of a complete icdl w/t-f $(A, \leq, N, T)$ iff $(A', \leq, N)$ is a complete sub-icdl of $(A, \leq, N)$ and $T' = A' \cap T$.

Let us define a *homomorphism* between De Morgan lattices $(A, \leq, N)$ and $(A', \leq', N')$ as a lattice homomorphism $h^{43}\|_{44}$ that has the further property that $h(Na) = N'h(a)$. A homomorphism between icdl's is then taken to be but a special case of this definition. We note the fact that the condition that

---

[18]They actually require that a *quasi-Boolean algebra* have a 0 and a 1, but since any lattice may be trivially so bounded, we neglect this distinction between De Morgan lattices and quasi-Boolean algebras in the sequel.

h preserves join follows from its preserving meet and De Morgan complementation. An *isomorphism* is a one-to-one homomorphism. We define a *T-preserving homomorphism*, or *T-homomorphism* between two icdl's w/t-f $(A, \leq, \mathrm{N}, T)$ and $(A', \leq', \mathrm{N}', T')$ as a homomorphism such that if $a \in T$, then $h(a) \in T'$. A *T-isomorphism* is a T-homomorphism that is an isomorphism. Let us note that the homomorphic image of a De Morgan lattice is a De Morgan sublattice of the target De Morgan lattice, the homomorphic image of an icdl is a sub-icdl of the target icdl, and the T-homomorphic image of an icdl w/t-f is a sub-icdl w/t-f of the target icdl w/t-f.

We may then define a *complete homomorphism* between complete De Morgan lattices as a homomorphism that is at the same time a complete lattice homomorphism. The condition that generalized join be preserved actually follows from the conditions that generalized meet and De Morgan complementation be preserved. A *complete isomorphism* is then a one-to-one complete homomorphism. A *complete T-homomorphism* between two complete icdl's w/t-f is then a T-homomorphism that is at the same time a complete homomorphism. A *complete T-isomorphism* is then a complete T-homomorphism that is an isomorphism. Note that the complete homomorphic image of a complete De Morgan lattice is a ⁴⁴‖₄₅ complete De Morgan sublattice of the target complete De Morgan lattice, the complete homomorphic image of a complete icdl is a complete sub-icdl of the target complete icdl, and the complete T-homomorphic image of a complete icdl w/t-f is a complete sub-icdl w/t-f of the target complete icdl w/t-f.

A *direct product* of De Morgan lattices is defined as follows: Let $\{(A_x, \leq_x, \mathrm{N}_x)\}_{x \in X}$ be an indexed set of De Morgan lattices, and let the *direct product* $\prod_{x \in X} A_x$ be that triple $(A, \leq, \mathrm{N})$ such that $A = \times_{x \in X} A_x$, $\{a_x\}_{x \in X} \leq \{b_x\}_{x \in X}$ iff $a_x \leq_x b_x$ for all $x \in X$, and $\mathrm{N}\{a_x\}_{x \in X} = \{\mathrm{N}_x a_x\}_{x \in X}$. Observe that the direct product of De Morgan lattices is also a De Morgan lattice, and that the direct product of complete De Morgan lattices is also a complete De Morgan lattice.

Belnap 1965 defined a notion of a *product* of icdl's w/t-f that is akin to the notion of a direct product, but subtly different. Let $\{(A_x, \leq_x, \mathrm{N}_x, T_x)\}_{x \in X}$ be an indexed set of icdl's w/t-f, and let the *product* $\prod_{x \in X} A_x$ be that quadruple $(A, \leq, \mathrm{N}, T)$ such that $T = \times_{x \in X} T_x$, $\overline{T} = \times_{x \in X} \overline{T}_x$, $A = T \cup \overline{T}$, $\{a_x\}_{x \in X} \leq \{b_x\}_{x \in X}$ iff $a_x \leq_x b_x$ for all $x \in X$, and $\mathrm{N}\{a_x\}_{x \in X} = \{\mathrm{N}_x a_x\}_{x \in X}$. Observe that the product of icdl's w/t-f is an icdl w/t-f, and that the product of complete icdl's w/t-f is a complete icdl w/t-f.

Belnap 1965 then showed utilizing proof-theoretical (syntactical) methods that every at most denumerable icdl w/t-f is T-isomorphic to a sublattice of some product of an at most denumerable collection of replicas of M_0. In ⁴⁵‖₄₆ the next chapter we shall give an algebraic proof that at the same time both generalizes and sharpens this result.

We mention in closing that Belnap 1965 showed that a *first degree formula* (a truth function of first degree entailments and pure truth functions) is a theorem of E (or R) iff it is valid in every finite product of M_0. We have as yet no interesting algebraic proof of this result. [46] $\|_{47}$

II. Homomorphisms of Intensionally Complemented Distributive Lattices

1. Prime Filters and Homomorphisms

We pointed out in section 5 of the first chapter that the icdl M_0 has a fundamental role in the semantics of E and R that is comparable to the fundamental role of the Boolean algebra **2** in the semantics of the classical propositional calculus. In this chapter we demonstrate that M_0 plays a similarly fundamental role in the algebra of icdl's that is comparable to the fundamental role of **2** in Boolean algebra.[1] Thus, just as prime filters and homomorphisms into **2** are connected in Boolean algebras, the following two theorems show a similar connection between prime filters and homomorphisms into M_0 in icdl's.

Theorem 1. *Every prime filter P of an icdl w/t-f $(A, \leq, \mathrm{N}, T)$ determines a T-homomorphism h of A into M_0 satisfying the following conditions (where F_i is the principal filter generated by i in M_0):*

i) $h(a) \in F_{-1}$ iff $a \in P$,

ii) $h(a) \in F_{-2}$ iff $\mathrm{N}a \in \overline{P}$,

iii) $h(a) \in F_{+0}$ iff $a \in T$.

It is clear that these conditions uniquely determine a (single-valued) function h, for if we let εF_i be either $^{47}\|_{48}$ F_i or $\overline{F}_i$, the set theoretic intersection of εF_{-1}, εF_{-2}, and εF_{+0} is always a unit set, for every such choice function ε.

We now prove that h is a T-homomorphism. We first show that $h(a \wedge b) = h(a) \wedge h(b)$. In view of our remarks above concerning the unique determination of h, it suffices to show that $h(a \wedge b) \in F_i$ iff $h(a) \wedge h(b) \in F_i$ $(i = -1, -2, +0)$. The cases $i = -1$ and $i = +0$ may be treated together. Thus $h(a \wedge b) \in F_{-1}\,[F_{+0}]$, iff $a \wedge b \in P\,[T]$, iff $a, b \in P\,[T]$, iff $h(a), h(b) \in F_{-1}\,[F_{+0}]$, iff $h(a) \wedge h(b) \in F_{-1}\,[F_{+0}]$. We treat $i = -2$ by using the fact that $\overline{P}$ is a prime ideal (a well-known consequence of the fact that P is a prime filter), together with an immediate De Morgan property of N. Thus $h(a \wedge b) \in F_{-2}$, iff $\mathrm{N}(a \wedge b) = \mathrm{N}a \vee \mathrm{N}b \in \overline{P}$, iff $\mathrm{N}a, \mathrm{N}b \in \overline{P}$, iff $h(a), h(b) \in F_{-2}$, iff $h(a) \wedge h(b) \in F_{-2}$.

---

[1] The essentials of this chapter appear in Dunn and Belnap 1965.

We now show that $h(\mathrm{N}a) = \mathrm{N}(ha)$. Again it suffices to show that $h(\mathrm{N}a) \in F_i$ iff $\mathrm{N}(ha) \in F_i$ $(i = -1, -2, +0)$. We first define $\mathrm{N}(F_i)$, as the set of $\mathrm{N}b$ such that $b \in F_i$, and observe that $F_{-1} = \mathrm{N}(\overline{F_{-2}}), F_{-2} = \mathrm{N}(\overline{F_{-1}})$, and $F_{+0} = \mathrm{N}(\overline{F_{+0}})$. Then $h(\mathrm{N}a) \in F_{-1}$, iff $\mathrm{N}a \in P$, iff $h(a) \in \overline{F_{-2}}$, iff $\mathrm{N}(ha) \in \mathrm{N}(\overline{F_{-2}}) = F_{-1}$. And $h(\mathrm{N}a) \in F_{-2}$, iff $\mathrm{NN}a = a \in \overline{P}$, iff $h(a) \in \overline{F_{-1}}$, iff $\mathrm{N}(ha) \in \mathrm{N}(\overline{F_{-1}}) = F_{-2}$. And finally, $h(\mathrm{N}a) \in F_{+0}$ iff $\mathrm{N}a \in T$, iff $a \in \overline{T}$, iff $h(a) \in \overline{F_{+0}}$ iff $\mathrm{N}(ha) \in \mathrm{N}(\overline{F_{+0}}) = F_{+0}$.

We complete our proof by noting that condition iii) insures that the homomorphism h is T-preserving.

We also have the converse theorem. [48] $\|_{49}$

Theorem 2. *Every T-homomorphism h of an icdl w/t-f $(A, \leq, \mathrm{N}, T)$ into M_0 determines a prime filter P of $(A, \leq, \mathrm{N}, T)$ in accord with conditions i) and ii) of Theorem 1.*

We first remark that, given a T-homomorphism h, condition i) by itself determines the set P, so it suffices to show that P is a prime filter and that it satisfies condition ii). That the set P is a prime filter follows immediately from the easily verified fact that if h is a homomorphism into a prime filter (in this case F_{-1}), then the inverse image of that prime filter under h (in this case P) is also a prime filter. We now demonstrate that P satisfies condition ii). Observing that $F_{-2} = \mathrm{N}(\overline{F_{-1}})$, we have that $h(a) \in F_{-2}$, iff $h(a) \in \mathrm{N}(\overline{F_{-1}})$, iff $\mathrm{N}(ha) = h(\mathrm{N}a) \in \mathrm{NN}(\overline{F_{-1}}) = \overline{F_{-1}}$. But contraposing condition i), we have $h(\mathrm{N}a) \in \overline{F_{-1}}$ iff $\mathrm{N}a \in \overline{P}$. So any set P which satisfies condition i) also satisfies condition ii), and our proof is complete.

We remark that by combining Theorems 1 and 2 we obtain a natural one-to-one correspondence between prime filters and T-homomorphisms into M_0. That no two different prime filters determine the same T-homomorphism is immediate, since they would differ at an element a, and hence the homomorphisms determined by them would differ in that one would send a into F_{-1} and the other would not. That no two different T-homomorphisms h and h' determine the same prime filter is shown as follows. Suppose for some [49] $\|_{50}$ element a, $h(a) \neq h'(a)$. We shall show that there is always some element b such that $h(b) \in F_{-1}$ but $h'(b) \notin F_{-1}$. The proof is by cases. Assume first that $h(a), h'(a) \in F_{-1}$. Make the further assumption that $a \in T$. Then since h and h' preserve T, $h(a), h'(a) \in \{+1, +3\}$. Assume without loss of generality that $h(a) = +1$ and $h'(a) = +3$. Then $h(\mathrm{N}a) = -1 \in F_{-1}$, but $h'(\mathrm{N}a) = -3 \notin F_{-1}$. On the assumption that $a \notin T$, it follows immediately from the fact that h and h' are T-preserving, that $h(a), h'(a) \in \{-1, -0\}$. Assume without loss of generality that $h(a) = -1$ and $h'(a) = -0$. Then $h(\mathrm{N}a) = +1 \in F_{-1}$, but $h'(\mathrm{N}a) = +0 \notin F_{-1}$. A glance at the diagram of M_0 should convince the reader that the remaining case, in which

$h(a), h'(a) \notin F_{-1}$, may also be divided according as to whether a is or is not in T, and treated analogously.

We obtain as an immediate consequence of Theorem 1, together with Stone's 1937 Prime Filter Separation Theorem for distributive lattices (cf. Ch. I, Sec. 3.), the following:

Theorem 3. *In an icdl w/t-f, for elements a and b such that $a \not\leq b$, there exists a T-homomorphism h of the icdl w/t-f into M_0 such that $h(a) \in F_{-1}$ and $h(b) \in \overline{F_{-1}}$, which means that $h(a) \neq h(b)$.*

Our proof of Theorem 3 implicitly uses the axiom of choice since Stone's proof uses the axiom of choice. However, the following may be proven without appealing to the axiom of choice (primarily because it is built into the $^{50}\|_{51}$ notion of an icdl that it must have a truth filter).

Theorem 4. *A De Morgan lattice is an icdl iff it has a homomorphism into M_0.*

It is thus an immediate corollary of Theorem 1 that for an arbitrary De Morgan lattice, the condition that it be an icdl, i.e., that it have a truth filter T, implies that have a homomorphism into M_0 (letting T play a double role, both as the truth filter and as the prime filter). To prove the converse, let $(A, \leq, N)$ be a De Morgan lattice and h a homomorphism of it into M_0. Let T be the inverse image of F_{+0} under h. As has already been remarked, the inverse image of a prime filter under a homomorphism is a prime filter. So it only remains to prove that for all $a \in A$, T contains exactly one of a and Na. Suppose that both $a, Na \in T$. Then $h(a)$ and $h(Na) = N(ha)$ are both members of F_{+0}, but this is impossible since F_{+0} is a truth filter. Suppose that for some a, $a, Na \notin T$. Then neither $h(a)$ nor $h(Na) = N(ha)$ are members of F_{+0}, but again this is impossible since F_{+0} is a truth filter.

2.   Prime Filters and Embeddings

As we promised in section 5 of the first chapter, we give an algebraic proof of a theorem that is both a generalization and a sharpening of a result of Belnap 1965. Following Belnap, we define for any cardinal c, $M^c = \prod_{x:\, x<c} M_{0_x} \,^{51}\|_{52}$ (a *product* of M_0, in the sense of section 5 of the first chapter). We prove

Theorem 1. *Every icdl w/t-f of cardinality d is T-isomorphic to a sub-icdl w/t-f of M^c, for some cardinal $c \leq d^2$.*

The proof begins by taking the set of all pairs of elements (a, b) of the given icdl w/t-f A such that $a \not\leq b$, and indexing these pairs with some set X.

Note that the cardinality of X, say c, is obviously such that $c \leq d^2$ where d is the cardinality of A. By Theorem 1.3, we know that with each pair $(a, b)_x$ we may associate a T-homomorphism h_x of A into M_0 so $h_x(a) \neq h_x(b)$.

Then h, defined for $a \in A$ by $h(a) = \{h_x(a)\}_{x \in X}$, is a T-isomorphism of A into M^c. The function h is obviously defined over all of A, since that is true for each h_x. And h is into M^c, and is T-preserving, because each h_x is into M_0 and T-preserving. And h is one-to-one, for suppose that $a \neq b$, and assume without loss of generality that $a \nleq b$. Then the T-homomorphism h_x associated with $(a, b)_x$ is such that $h_x(a) \neq h_x(b)$, and $h(a)$ differs from $h(b)$ in the x-th component.

It remains to show h a homomorphism; this follows immediately from the fact that operations in the product algebra were defined component-wise: $h(a \wedge b) = \{h_x(a \wedge b)\}_{x \in X} = \{h_x(a) \wedge h_x(b)\}_{x \in X} = h(a) \wedge h(b)$, and $h(\mathrm{N}a) = \{h_x(\mathrm{N}a)\}_{x \in X} = \{\mathrm{N}h_x(a)\}_{x \in X} = \mathrm{N}(ha)$. [52]$\|_{53}$

We mention that corresponding to this embedding theorem we have a sort of representation theorem in terms of a ring of sets with an intensional complement N. This theorem, however is not nearly as interesting as the representation theorem for Boolean algebras that corresponds to the embedding theorem of Boolean algebras into the product algebra of the two-element Boolean algebra, since the intensional complement on the ring of sets is not as natural an operation as might be desired. The representation for icdl's w/t-f follows trivially from the fact that every distributive lattice is isomorphic to a ring of sets under the mapping h which associates with each element a the set of all prime filters P such that $a \in P$ (cf. chapter I, section 3). Defining $\mathrm{N}(ha)$ to be the set of all prime filters P such that $\mathrm{N}a \in P$, we trivially insure that h preserves N. Perhaps the only interest such a representation theorem has derives from the demonstrated one-to-one correspondence between prime filters and T-homomorphisms into M_0. From this it is easy to see that the mapping h which associates with each element a the set of T-homomorphisms that carry a into F_{-1} is an iso-morphism onto a ring of sets that preserves intensional complementation when $\mathrm{N}(ha)$ is defined as the set of T-homomorphisms that carry a into the ideal generated by -1, which is just $\mathrm{N}(F_{-1})$. We shall establish a more interesting representation in Theorem 4 of chapter V. [53]$\|_{54}$

3. Complete, Completely Prime Filters and Complete Homomorphisms

We prove the following complete generalization of Theorem 1.1:

Theorem 1. *Every complete and completely prime filter P of a complete icdl w/t-f $(A, \leq, \mathrm{N}, T)$ determines a complete T-homomorphism h of A into*

M_0 that satisfies conditions i) ii), and iii) of Theorem 1.1.

In view of Theorem 1.1 and the remarks concerning the strategy of its proof, it suffices to show that $h(\bigwedge_{x \in X} a_x) \in F_i$ iff $\bigwedge_{x \in X} h(a_x) \in F_i$ ($i = -1, -2, +0$). Again we treat $i = -1$ and $i = +0$ together, observing first that F_{-1} [F_{+0}] is complete (since M_0 is finite) and that P [T] is complete. Thus, $h(\bigwedge_{x \in X} a_x) \in F_{-1}$ [F_{+0}], iff $\bigwedge_{x \in X} a_x \in P$ [T], iff $\{a_x\}_{x \in X} \subseteq P$ [T], iff $\{h(a_x)\}_{x \in X} \subseteq F_{-1}$ [F_{+0}], iff $\bigwedge_{x \in X} h(a_x) \in F_{-1}$ [F_{+0}]. We treat $i = -2$ by utilizing the fact that $\overline{P}$ is a complete and completely prime ideal, together with a De Morgan property of N over $\bigwedge$. Thus, $h(\bigwedge_{x \in X} a_x) \in F_{-2}$, iff $N \bigwedge_{x \in X} a_x = \bigvee_{x \in X} Na_x \in \overline{P}$ iff $\{Na_x\}_{x \in X} \subseteq \overline{P}$, iff $\{h(a_x)\}_{x \in X} \subseteq F_{-2}$, iff $\bigwedge_{x \in X} h(a_x) \in F_{-2}$.

We again also have the converse theorem:

Theorem 2. *Every complete T-homomorphism h of a complete icdl w/t-f $(A, \leq, N, T)$ into M_0 determines a complete and completely prime filter P of $(A, \leq, N, T)$ in accord with conditions i) and ii) of Theorem 1.1.* [54]‖[55]

That condition i) determines P to be a complete and completely prime filter follows immediately from the easily verified fact that if h is a complete homomorphism into a complete and completely prime filter (in this case F_{-1}), then the inverse image of that complete and completely prime filter under h (in this case P) is a complete and completely prime filter also. The proof that P as determined by i) also satisfies ii) is the same as in Theorem 1.2.

We remark that by combining Theorems 1 and 2 we obtain a natural one-to-one correspondence between complete, completely prime filters and complete T-homomorphisms into M_0. The justification of this remark rests on the same argument as after Theorem 1.1.

At this point a complete generalization of Theorem 1.3 would be appropriate, showing that distinct elements of a complete icdl w/t-f may be separated by a complete T-homomorphism into M_0. But such a theorem cannot be proven without restriction, for we are not guaranteed the key assumption that for any two elements a, b such that $a \not\leq b$, there exists a complete and completely prime filter P such that $a \in P$ and $b \notin P$. We find that this assumption is unjustified in the case of complete icdl's w/t-f. Thus, let A be the union of the two closed intervals $[0, 1/3]$ and $[2/3, 1]$ of the real line with the usual ordering. It is easily seen that this is a complete and completely distributive lattice, and that if Na is defined as $1 - a$, then [55]‖[56] N is antitone and of period two, and $[2/3, 1]$ becomes a complete truth filter. It is easy to see that there is no complete and completely prime filter that is a subset of the half-open interval $(2/3, 1]$, for suppose there is such a filter P. Since P is complete, for some $a \in (2/3, 1]$, P is the principal filter generated by a. But then consider $\overline{P} = \{b \in A : b < a\}$. It is well-known from analysis that $a = \bigvee \overline{P}$, but since P and $\overline{P}$ are disjoint, P is not completely prime. Thus, no two points in $(2/3, 1]$ can be separated by a complete and completely prime filter.

In the next section we investigate necessary and sufficient conditions for the separation of distinct elements by complete, completely prime filters, at the same time determining the range of applicability of a complete generalization of the embedding Theorem 2.1.

4. Complete, Completely Prime Filters and Embeddings

If we assume that a complete icdl w/t-f is such that any two distinct elements can be separated by a complete and completely prime filter, then we may prove a complete generalization of the embedding Theorem 2.1. With the obvious association of a complete T-homomorphism h_x into M_0 with each pair $(a, b)_x$, the proof goes through just as before. Indeed, the associated complete T-homomorphisms would then define a T-isomorphism h into M^c just as in Theorem 2.1. And obviously h preserves $\bigwedge$, since each $h_x\,^{56}\|_{57}$ preserves $\bigwedge$, and $\bigwedge$ in M^c is defined componentwise. Thus we have

Lemma 3. *Every complete icdl w/t-f $(A, \leq, N, T)$ of cardinality d, and such that for every pair of elements a, b with $a \not\leq b$, there exists a complete and completely prime filter P such that $a \in P$ and $b \notin P$, is completely T-isomorphic to a complete sublattice of M^c for some cardinal $c < d^2$.*

We also have a converse form of Lemma 1, namely,

Lemma 4. *Every complete icdl w/t-f $(A, \leq, N, T)$ which is completely T-isomorphic to a complete sublattice of M^c for some cardinal c, is such that for every pair of elements a, b with $a \not\leq b$, there exists a complete and completely prime filter P such that $a \in P$ and $b \notin P$.*

In view of the given T-isomorphism of A onto a complete sublattice A' of M^c, it suffices to show that for $a', b' \in A'$ with $a' \not\leq b'$, there exists a complete and completely prime filter P' such that $a' \in P'$ and $b' \notin P'$. And to show this, it suffices to show that for $\{a_x\}_{x<c}, \{b_x\}_{x<c} \in M^c$ with $\{a_x\}_{x<c} \not\leq \{b_x\}_{x<c}$, there exists a complete and completely prime filter Q of M^c such that $\{a_x\}_{x<c} \in Q$ and $\{b_x\}_{x<c} \notin Q$. For let Q be a complete and completely prime filter of M^c such that $a' \in Q$ and $b' \notin Q$. Then it is easy to see that $Q \cap A'$ is a complete and completely prime filter of A' such that $a' \in Q \cap A'$ and $b' \notin Q \cap A'$. So consider $\{a_x\}_{x<c}, \{b_x\}_{x<c} \in M^c$ with $\{a_x\}_{x<c} \not\leq \{b_x\}_{x<c}$ Then for some $x_0 < c$, $a_{x_0} \not\leq b_{x_0}$. Then there exists a complete and completely prime filter Q_x of $^{57}\|_{58}$ M_{x_0} such that $a_{x_0} \in Q_{x_0}$ and $b_{x_0} \notin Q_{x_0}$, as inspection of M_0 shows. Now define functions f and g on $\{x : x < c\}$ such that $f(x_0) = Q_{x_0} \cap F_{+0}$ and for $x \neq x_0$, $g(x) = \overline{F_{+0}}$, and such that $g(x_0) = Q_{x_0} \cap \overline{F_{+0}}$ and for $x \neq x_0$, $f(x) = F_{+0}$. Then let $Q = \times_{x<c} f(x) \cup \times_{x<c} g(x)$. Componentwise considerations now show that Q is the desired complete and completely prime filter of M^c.

It is interesting to note that the condition that every pair of distinct elements be separated by a complete and completely prime filter is equivalent to a rather simple condition on the elements.

Lemma 5. *Given a complete and completely distributive lattice $(A, \leq)$, a necessary and sufficient condition for any pair of distinct elements being separated by a complete and completely prime filter is that every element be a generalized join of completely join-irreducible elements, where an element a is said to be completely join-irreducible if for no set $\{a_x\}_{x \in X}$ such that each $a_x < a$, does $a = \bigvee_{x \in X} a_x$.*

The sufficiency is easy, for suppose that $a \not\leq b$, and say $a = \bigvee_{x \in X} a_x$, where each a_x is completely join-irreducible. Now choose some $a_x \not\leq b$. (There must be one, for if all $a_x \leq b$, then $a = \bigvee_{x \in X} a_x \leq b$, contrary to assumption.) Since $a_x \not\leq b$, the principal filter P generated by that a_x contains a, but does not contain b. And it is trivial that $^{58}\|_{59}$ P is complete, as are all principal filters in a complete lattice. It only remains to show P completely prime. But this follows immediately from a complete generalization of a lemma in Birkhoff 1948, namely that in a complete and completely distributive lattice, if an element p is completely join-irreducible, then $p \leq \bigvee_{x \in X} q_x$ implies $p \leq q_x$ for some $x \in X$. For proof of the generalization, note that $p \leq_{x \in X} q_x$ implies $p = p \wedge (\bigvee_{x \in X} q_x) = \bigvee_{x \in X} p \wedge q_x$. But then since p is completely join-irreducible, $p = p \wedge q_x$, i.e., $p \leq q_x$, for some $x \in X$.

The necessity follows by letting $\{p_x\}_{x \in X}$ be the set of all completely join-irreducible elements p_x such that $p_x \leq a$ for some arbitrary element a. This set is non-empty since $\bigwedge A$ is completely join-irreducible. Now if $a = \bigvee_{x \in X} p_x$ we are through. So assume $a \not\leq \bigvee_{x \in X} p_x$. But then there is by assumption a complete and completely prime filter P that contains a and does not contain $\bigvee_{x \in X} p_x$. But $\bigwedge P$ is completely join-irreducible, for otherwise P, being completely prime, would contain some element $q < \bigwedge P$. Thus $\bigwedge P = p_x$ for some $x \in X$, and $\bigwedge P \leq \bigvee_{x \in X} p_x$, contrary to the hypothesis that P does not contain $\bigvee_{x \in X} p_x$.

We remark in passing that Lemma 3, although handy for our purposes, is not stated in the best possible form, for in its proof we make use of the condition of complete distributivity only in the proof of the complete generalization of Birkhoff's lemma, and there only in a weakened form. Further we could do without any condition on [59][60] distributivity by defining a *completely join-prime* element as one satisfying the consequent of the generalization of Birkhoff's lemma, and formulating the necessary and sufficient condition of Lemma 3 in terms of completely-join-primeness instead of completely-join-irreducibility.

We now collect our lemmas together in the following theorem:

Theorem 1. *For a complete icdl w/t-f $(A, \leq, N, T)$, the following conditions are equivalent:*

(i) *There exists a complete T-isomorphism of A into M^c, for some cardinal c.*

(ii) *For any pair of elements a, b with $a \nleq b$ there exists a complete and completely prime filter P such that $a \in P$ and $b \notin P$.*

(iii) *Every element a is a generalized join of completely join-irreducible elements.*

We remark in closing then that for a complete icdl w/t-f that satisfies one of these conditions i)-iii), we get the same sort of trivial representation theorem as was described in section 2, this time mapping each element into the set of complete and completely prime filters that contain it, or into the set of complete T-homomorphisms that carry it into F_{+0} of M_0.

5.   Complete Icdl's and Complete De Morgan Lattices

Now, as promised in section 5 of the first chapter, we prove a complete generalization of the theorem of Belnap [60][61] and Spencer 1964 that a necessary and sufficient condition for a De Morgan lattice to be an icdl is that N have no fixed point. The proof that we are about to give is entirely due to Nuel D. Belnap, Jr., who communicated it to the author. But its detailed presentation here is due to the author, who hence takes responsibility for any mistakes.

Theorem 1. *A necessary and sufficient condition for a complete De Morgan lattice $(A, \leq, N)$ to be a complete icdl is that for all $a \in A$, $a \neq Na$.*

The necessity is obvious, because for all $a \in A$, T must contain exactly one of a and Na.

To prove sufficiency, we begin by defining Σ to be the collection of all selection sets S, where S is a selection set iff for every $a \in A$, S contains at least one of a and Na. To prove that A includes a complete truth filter T, it then suffices to show that for some $S \in \Sigma$, $\bigwedge S \nleq N \bigwedge S$. For let

S_0 be such an S, and let T be the principal filter generated by $\bigwedge S_0$. It is immediate that T, being principal, is complete. And T trivially contains for every $a \in A$, at least one of a and Na. Further, T contains at most one of a and Na, for suppose to the contrary that $\bigwedge S_0 \leq a$ and $\bigwedge S_0 \leq Na$. Since S_0 contains one of a and Na. Thus $\bigwedge S_0 \leq \bigvee NS_0$, but $\bigvee NS_0 = N \bigwedge S_0$, and thus $\bigwedge S_0 \leq N \bigwedge S_0$, contrary to hypothesis.

The proof strategy then is to suppose for *reductio* that there is no such selection set S_0, i.e., that for all [61]$\|_{62}$ $S \in \Sigma$, $\bigwedge S \leq N \bigwedge S$, and to derive under this hypothesis that N has a fixed point.

First well-order the elements of A, and let $a \prec b$ mean that a properly precedes b in the well-ordering. For $x \in A$, let $F(x)$ be $\{x \wedge Nx\} \cup \{y \vee Ny \colon y \in A \,\&\, y \prec x\}$, and let $G(x)$ be $\{x \vee Nx\} \cup \{y \wedge Ny \colon y \in A \,\&\, y \prec x\}$. Then define $a_0 = \bigvee_{x \in A} \bigwedge F(x)$, whence $Na_0 = \bigwedge_{x \in A} \bigvee G(x)$. We show that $a_0 = Na_0$.

To prove that $a_0 \leq Na_0$ reduces to showing for arbitrary elements p and q, that $\bigwedge F(p) \leq \bigvee G(q)$. We have three cases according to the relative position of p and q in the well-ordering. If $p = q$, then $p \wedge Np \in F(p)$ and $p \vee Np \in G(q)$, but $p \wedge Np \leq p \vee Np$. If $p \prec q$, then $p \wedge Np \in F(p)$ and $p \wedge Np \in G(q)$. And if $q \prec p$, then $q \vee Nq \in F(p)$ and $q \vee Nq \in G(q)$.

To show that $Na_0 \leq a_0$ we use complete distribution. Thereby we find that $a_0 = \bigwedge_{f \in U} \bigvee_{x \in A} f(x)$, where U is the set of all functions f of A such that $f(x) \in F(x)$. And $Na_0 = \bigvee_{g \in V} \bigwedge_{x \in A} g(x)$, where V is the set of all functions g of A such that $g(x) \in G(x)$. To prove that $Na_0 \leq a_0$, it suffices to show for some arbitrary $g \in V$ and $f \in U$, that $\bigwedge_{x \in A} g(x) \leq \bigvee_{x \in A} f(x)$. Proof is by cases. First suppose that for some $y \in A$, there exists $p \in A$ such that $g(p) = y \wedge Ny$, and that for some $z \in A$, there exists $q \in A$ such that $f(q) = z \vee Nz$. Invoking the well-ordering principle, let y_0 be the first such element y, and let z_0 be the first such element z. If $y_0 = z_0$, then $y_0 \wedge Ny_0 \in \{g(x)\}_{x \in A}$ and $y_0 \vee Ny_0 \in \{f(x)\}_{x \in A}$, but $y_0 \wedge Ny_0 \leq_0 y_0 \vee Ny_0$. If $y_0 \prec z_0$, then either $f(y_0) = y_0 \wedge Ny_0$, in which case $y_0 \wedge Ny_0$ is [62]$\|_{63}$ contained in both $\{g(x)\}_{x \in A}$ and $\{f(x)\}_{x \in A}$, or $f(y_0) = y \vee Ny$ for some $y \prec y_0$. But this last is impossible, for then $y \prec y_0 \prec z_0$, contradicting that z_0 is first among the elements z such that for some $q \in A$, $f(q) = z \vee Nz$. An exactly parallel argument holds for the supposition that $z_0 \prec y_0$.

Next suppose that for all $p \in A$, $g(p) = p \vee Np$, and that for some $q \in A$, $f(q) = z \vee Nz$ for some $z \prec q$. Then for that z, $g(x) = z \vee Nz$, and thus $z \vee Nz$ is in both $\{g(x)\}_{x \in A}$ and $\{f(x)\}_{x \in A}$.

Finally, suppose that for all $p \in A$, $g(p) = p \vee Np$, and for all $q \in A$, $f(q) = q \wedge Nq$. Then $\{g(x)\}_{x \in A} = \{x \vee Nx\}_{x \in A}$, and $\{f(x)\}_{x \in A} = \{x \wedge Nx\}_{x \in A}$, so we must show $\bigwedge_{x \in A} x \vee Nx \leq \bigvee_{x \in A} x \wedge Nx$. But this follows by complete distribution from $\bigvee_{S \in \Sigma} \bigwedge S \leq \bigwedge_{S \in \Sigma} \bigvee S$, which in turn follows from our *reductio* supposition that for all $S \in \Sigma$, $\bigwedge S \leq N \bigwedge S$. Thus, consider arbitrary $S_0, S_1 \in \Sigma$. If S_0 and S_1 share some element, then $\bigwedge S_0 \leq \bigvee S_1$. Otherwise, $S_1 = \mathrm{N} S_0$, and by our reductio assumption $\bigwedge S_0 \leq \mathrm{N} \bigwedge S_0 = \bigvee \mathrm{N} S_0 = \bigvee S_1$.

Then $a_0 = \mathrm{N} a_0$, contrary to the hypothesis that N has no fixed point; and so for some $S \in \Sigma$, $\bigwedge S$ must generate a complete truth filter.

We also have a complete generalization of Theorem 1.4.

Theorem 2. *A necessary and sufficient condition for a complete De Morgan lattice* $(A, \leq, \mathrm{N})$ *to be a complete icdl is that there be a complete homomorphism* h *of* A *into* M_0.

The necessity of this condition is an immediate corollary of Theorem 3.1, with T playing the double role of $^{63}\|_{64}$ the truth filter and the complete and completely prime filter. The sufficiency follows by letting T be the inverse image of F_{+0} under the assumed complete homomorphism h, and noting, as has already been remarked, that the inverse image of a complete, completely prime filter under a complete homomorphism is itself a complete, completely prime filter. Thus T so defined is complete, and that T is a truth filter follows from the proof of Theorem 1.4. $^{64}\|_{65}$

III. Homomorphisms of De Morgan Lattices

1. Prime Filters and Homomorphisms

Kalman 1958 showed that every De Morgan lattice is embeddable into a direct product of a certain De Morgan lattice which we shall call D, and which has the following diagram, where $N(1) = 0$ (hence $N(0) = 1$), $N(p) = p$, and $N(q) = q$:

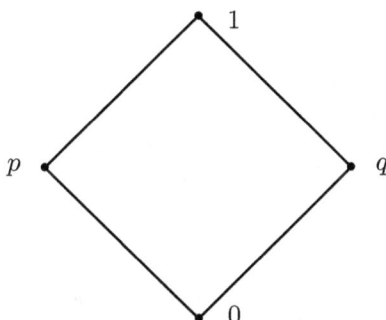

As is suggested by Kalman's result, we shall demonstrate that D has a fundamental role in the study of De Morgan lattices similar to that of M_0 among icdl's and that of **2** among Boolean algebras. Indeed, in this chapter we shall prove Kalman's embedding theorem by developing much the same connections between prime filters and homomorphisms as we found in chapter II for icdl's. This may be heuristically illuminating insofar as Kalman's proof comes very quickly from a high-powered theorem of universal $^{65}\|_{66}$ algebra and does not itself suggest the connections between prime filters and homomorphisms into D that we are about to develop. Our present proof, though not as streamlined as Kalman's, moves more slowly and thus allows us to see more of the neighboring countryside.

Thus corresponding to Theorem II:1.1, we have

Theorem 1. *Every prime filter P of a De Morgan lattice $(A, \leq, N)$ determines a homomorphism h of A into D satisfying the following conditions (where F_i means the principal filter generated by the element i in D):*

i) *$h(a) \in F_p$ iff $a \in P$,*

ii) *$h(a) \in F_q$ iff $Na \in \overline{P}$.*

The proof runs along the same lines as the proof of Theorem II:1.1. First we observe that these conditions uniquely determine a (single-valued) function h, for if we let εF_i be either F_i or $\overline{F_i}$, the set theoretical intersection of εF_p and εF_q is always a unit set, for every such choice function ε.

We now show that $h(a \wedge b) = h(a) \wedge h(b)$. It suffices to show that $h(a \wedge b) \in F_i$ iff $h(a) \wedge h(b) \in F_i$ ($i = p, q$). The arguments rest upon immediate consequences of the fact that both F_p and F_q are prime filters. Thus treating the case $i = p$, $h(a \wedge b) \in F_p$, iff $a \wedge b \in P$, iff $a, b \in P$, iff $h(a), h(b) \in F_p$, iff $h(a) \wedge h(b) \in F_p$. In treating the case $i = q$, we utilize the fact that $\overline{P}$ is a prime ideal (a consequence of the fact that P is a prime filter), together with a De Morgan property of N. Thus $h(a \vee b) \in F_q$, iff $^{66}\|_{67}$ $\mathrm{N}(a \wedge b) = \mathrm{N}a \vee \mathrm{N}b \in \overline{P}$, iff $\mathrm{N}(a), \mathrm{N}(b) \in \overline{P}$, iff $h(a), h(b) \in F_q$, iff $h(a) \wedge h(b) \in F_q$.

We now show that $h(\mathrm{N}a) = \mathrm{N}(ha)$. Again it suffices to show that $h(\mathrm{N}a) \in F_i$ iff $\mathrm{N}(ha) \in F_i$ ($i = p, q$). We first define $\mathrm{N}(F_i)$ as the set of $\mathrm{N}b$ such that $b \in F_i$, and observe that $F_p = \mathrm{N}(\overline{F_q})$ and $F_q = \mathrm{N}(\overline{F_p})$. Then $h(\mathrm{N}a) \in F_p$, iff $\mathrm{N}a \in P$, iff $h(a) \in \overline{F_q}$, iff $\mathrm{N}(ha) \in \mathrm{N}(\overline{F_q}) = F_p$. And $h(\mathrm{N}a) \in F_q$, iff $\mathrm{NN}a = a \in \overline{P}$, iff $h(a) \in \overline{F_p}$, iff $\mathrm{N}(ha) \in \mathrm{N}(\overline{F_p}) = F_q$, which completes the proof.

We also have the converse theorem:

Theorem 2. *Every homomorphism h of a De Morgan lattice $(A, \leq, \mathrm{N})$ into D determines a prime filter P of $(A, \leq, \mathrm{N})$ in accord with conditions i) and ii) of Theorem 1.*

We first remark that, given such a homomorphism h, condition i) by itself determines the set P, so it suffices to show that P is a prime filter and that it satisfies condition ii). It is trivial that the inverse image of a prime filter under a homomorphism is also a prime filter, so since clearly F_p is prime, so is P. We now show that P satisfies condition ii). Observing that $F_q = \mathrm{N}(\overline{F_p})$, we have that $h(a) \in F_q$, iff $h(a) \in \mathrm{N}(\overline{F_p})$, iff $\mathrm{N}(ha) = h(\mathrm{N}a) \in \mathrm{NN}(\overline{F_p}) = \overline{F_p}$. But contraposing condition i), we have that $h(\mathrm{N}a) \in \overline{F_p}$ iff $\mathrm{N}a \in \overline{P}$. So we have that $h(a) \in F_q$ iff $\mathrm{N}a \in \overline{P}$, which completes our proof.

By combining Theorems 1 and 2 we obtain a natural one-to-one correspondence between prime filters and homomorphisms into D. It is immediate that no two different $^{67}\|_{68}$ prime filters determine the same homomorphism, since they would then differ at some element a, and hence the homomorphisms determined by them would differ in that one would send a into F_p and the other would not. That no two different homomorphisms h and h' determine the same prime filter is shown as follows. Suppose for some element a, $h(a) \neq h'(a)$. We shall show that there is always some element b such that $h(b) \in F_p$ but $h'(b) \notin F_p$, or vice versa. Assume first that $h(a), h'(a) \in F_p$. Assume without loss of generality that $h(a) = p$ and

$h'(a) = 1$. Then $h(\mathrm{N}a) = p \in F_p$, but $h'(\mathrm{N}a) = 0 \notin F_p$. Assume finally that $h(a), h'(a) \in \overline{F_p}$. Assume without loss of generality that $h(a) = q$ and that $h'(a) = 0$. Then $h(\mathrm{N}a) = q \notin F_p$ but $h'(\mathrm{N}a) = 1 \in F_p$.

We also have an analogue of Theorem II:1.3, namely,

Theorem 3. *In a De Morgan lattice, for elements a and b such that $a \nleq b$, there exists a homomorphism h of the De Morgan lattice into D such that $h(a) \in F_p$ and $h(b) \in \overline{F_p}$, which means that $h(a) \neq h(b)$.*

2. Prime Filters and Embeddings

Let us define for any cardinal c, $D^c = \prod_{x \,:\, x < c} D_x$ (a *direct* product of D). We are now in a position to give the proof promised in the proceeding section of a version of Kalman's embedding theorem, namely,

Theorem 1. *Every De Morgan lattice $(A, \leq, \mathrm{N})$ of cardinality d is isomorphic to a De Morgan sublattice $^{68}\|_{69}$ of D^c, for some cardinal $c \leq d^2$.*

Our proof is analogous to that of Theorem II:2.1. We take all pairs of elements (a, b) such that $a \nleq b$ and index them with some set X. Again we note that the cardinality of X, say c, is such that $c \leq d^2$, where d is the cardinality of A. By Theorem 1.3, we know that with each pair $(a, b)_x$ we may associate a homomorphism of A into D so $h_x(a) \neq h_x(b)$.

Then h, defined for $a \in A$ by $h(a) = \{h_x(a)\}_{x \in X}$, is an isomorphism of A into D^c, the proof of this being just like the proof of Theorem II:2.1.

Kalman 1958 proved this embedding theorem as a corollary of Birkhoff's 1944 theorem that every abstract algebra can be represented as a subdirect union of subdirectly irreducible homomorphic images of the algebra. Kalman observed that every homomorphic image of a De Morgan lattice is a De Morgan lattice, and that the only subdirectly irreducible De Morgan lattices are D and its subalgebras. Now this method of proof corresponds to the method in Birkhoff 1948 for showing that every Boolean algebra is embeddable in a product of the two element Boolean algebra by showing that (except for the degenerate Boolean algebra) the two element Boolean algebra is the only subdirectly irreducible Boolean algebra. On the other hand, our method of proof of the embedding theorem for De Morgan lattices corresponds to the well-known method of $^{69}\|_{70}$ proving the embedding theorem for Boolean algebras by utilizing the close connection between maximal filters and two-valued homomorphisms discovered by Stone 1936 (cf. Sikorski 1964 for a precise analogue to our method of proof).

3. Complete, Completely Prime Filters: Complete Homomorphisms; and Complete Embeddings

We also have the appropriate complete generalizations of the results of sections 1 and 2, corresponding to the complete generalizations of the results we obtained for icdl's. Thus we have the appropriate analogue of Theorem II:3.1, namely,

Theorem 1. *Every complete and completely prime filter P of a complete De Morgan lattice $(A, \leq, \mathrm{N})$ determines a complete homomorphism h of A into D that satisfies conditions i) and ii) of Theorem 1.1.*

The proof of this theorem is just an obvious modification of the proof of Theorem 1.1, just as the proof of Theorem II:3.1 was an obvious modification of the proof of Theorem II:1.1. The following converse theorem falls out in the same "copy-cat" fashion:

Theorem 2. *Every complete homomorphism h of a complete De Morgan lattice $(A, \leq, \mathrm{N})$ into D determines a complete and completely prime filter P of A in accord with conditions i) and ii) of Theorem 1.1.* $^{70}\|_{71}$

We also get the expected one-to-one correspondence between complete, completely prime filters and complete homomorphisms into D.

And we obtain the following analogue of Theorem II:4.1:

Theorem 3. *For a complete De Morgan lattice $(A, \leq, \mathrm{N})$, the following conditions are equivalent.*

   (i) *There exists a complete isomorphism of A into D^c, for some cardinal c.*

   (ii) *For any pair of elements a, b with $a \not\leq b$ there exists a complete and completely prime filter P such that $a \in P$ and $b \notin P$.*

   (iii) *Every element a is a generalized join of completely join-irreducible elements.* $^{71}\|_{72}$

IV. Simplicity and Normality

As one observes the special role of M_0 among icdl's w/t-f, one gets the idea that M_0 is somehow exemplary of icdl's w/t-f in general. One similarly gets the idea that D is exemplary of De Morgan lattices, and that 2 is exemplary of Boolean algebras. These ideas might be naturally expressed by saying that these structures are *simple*, but not so simple as to be *abnormal*. We shall now attempt to give these vague observations algebraic content.

1. Simplicity

It is well-known that the Boolean algebra 2 is *simple* in the sense of having only the two *trivial* congruences, i.e., the universal relation and the identity relation. It is also well-known that 2 is almost the only simple Boolean algebra (the degenerate Boolean algebra 1 being the only exception). We shall show something similar for D.

Let us define a congruence relation $\equiv$ on a De Morgan lattice $(A, \leq, \mathrm{N})$ as an equivalence relation on A with the following replacement properties:

If $a \equiv b$ and $c \equiv d$, then

(1) $a \wedge c \equiv b \wedge d$;

(2) $a \vee c \equiv b \vee d$; and

(3) $\mathrm{N}a \equiv \mathrm{N}b$. [72]$\|_{73}$

We observe that (1) follows from (2) and (3), and that because of the commutativity of $\vee$ we may replace (2) with a less complicated, condition (2'): $a \vee x \equiv b \vee x$ (as can be done for any lattice). These observations facilitate verification of congruences.

We recall as a part of universal algebra (cf. section 2 of chapter I) the mutual determination of congruences and homomorphisms. Descending now from universal algebra to the particularity of De Morgan lattices, it is easily verified that the quotient algebra of a De Morgan lattice is itself a De Morgan lattice. This means that everything is in order so that we may in the study of De Morgan lattices, as usual, reason interchangeably with their congruence relations and their homomorphisms.

Now defining a De Morgan lattice as *simple* iff it has only the two trivial congruences, we may show that D has a unique position among simple De Morgan lattices.

Theorem 1. *Up to isomorphism, all and only the De Morgan sublattices of D are simple De Morgan lattices.*

We shall verify by computation that every De Morgan sublattice of D is simple. We shall first verify that D itself is simple. The strategy will be to assume that two distinct elements of D are congruent and then to show under this assumption that all the elements of D are congruent. Let us first remark that if $0 \equiv 1$, then for every element a, $a = a \vee 0 \equiv a \vee 1 = 1$, and hence all elements are congruent to 1 and thus to each other. We shall reduce all other $^{73}\|_{74}$ cases to this case. Considerations of duality and symmetry convince us that it suffices to consider out of the remaining cases only the case where $p \equiv q$ and the case where $p \equiv 1$. If $p \equiv q$, then $p \wedge q \equiv q \wedge q$, and hence $0 \equiv q$. But also then $p \vee q \equiv q \vee q$, and hence $1 \equiv q$. From these two consequences we infer via the symmetry and transitivity of $\equiv$ that $0 \equiv 1$. On the other hand, if $p \equiv 1$, then $Np \equiv N1$, and hence $p \equiv 0$, which leads via the symmetry and transitivity of $\equiv$ to $0 \equiv 1$. So D is simple.

That the three element chain that is a De Morgan sublattice of D is simple may be seen by the same argument as in the last case above. Thus since the two element and the one element De Morgan sublattices of D are trivially simple, all De Morgan sublattices of D are simple.

To prove the "only" part of the theorem, we assume that A is a simple De Morgan lattice consisting of at least two distinct elements a and b (if A consists of just a single element it is isomorphic to that sub-De Morgan lattice of D which contains just p). Then by Theorem III:1.3, we know that there exists a homomorphism h of A into D such that $h(a) \neq h(b)$. Now either h is one-to-one, in which case A is isomorphic to a De Morgan sublattice of D, or h is not one-to-one. But then A has distinct elements c and d such that $h(c) = h(d)$. We then observe that the natural congruence $\equiv$ determined by h is not the identity relation since although $c \neq d$, $c \equiv d$. But neither is $\equiv$ the $^{74}\|_{75}$ universal relation, since $a \not\equiv b$. So unless h is an isomorphism, A is not simple, which completes the proof.

2. T-Simplicity

Since an icdl w/t-f is a special kind of De Morgan lattice, we might think it appropriate to define a congruence relation for icdl's w/t-f as we did for De Morgan lattices. But if we were to do so, we would be in trouble. The universal algebra that links up homomorphisms and congruences would still be all right (since it is *universal* algebra), but it would be inapplicable in the sense that it would not be true that the quotient algebra of an icdl w/t-f is an icdl w/t-f. As a simple example, consider that any icdl w/t-f would have the one element De Morgan lattice as a quotient algebra under the universal relation. Note, however, that the universal relation would relate elements of the truth filter with elements not in the truth filter. Logical motivations preclude such a monistic identification of truth and falsity. Algebraic motivations also count against such a congruence, for, since we studied icdl's w/t-f in terms of T-preserving homomorphisms in chapter II, it should be expected that the appropriate notion of congruence would be T-preserving as well.

Let us then define a T-congruence on an icdl w/t-f $(A, \leq, N, T)$ as a De Morgan congruence $\equiv$ (as defined in 1) [75]$\|_{76}$ that has the further property that $a \equiv b$ only if either $a, b \in T$ or $a, b \in \overline{T}$. It is now readily seen that T-homomorphisms determine T-congruences in the natural fashion. It is also readily seen that the quotient algebra of an icdl w/t-f $(A, \leq, N, T)$ under a T-congruence $\equiv$ is also an icdl w/t-f, the truth filter being the family of T-congruence classes of elements of T (which family we shall call the *natural truth filter*).

Now that we have the appropriate notion of congruence for icdl's w/t-f, what might be an appropriate notion of simplicity? Let us first note that it is easily verified for any icdl w/t-f that the relation which relates all members of the truth filter T to each other and all members of $\overline{T}$ to each other is a T-congruence. Let us call this congruence the *t-f congruence*. It does not seem out of line to call the t-f congruence a trivial congruence since it exists on every icdl w/t-f, and since it replaces the universal relation as the largest, least discriminating congruence. By regarding the t-f congruence as trivial, we may thus preserve the form and the spirit of the ordinary notion of simplicity with the following definition:

An icdl w/t-f is T-*simple* iff it has only the two trivial T-congruences, namely, the t-f congruence and the identity congruence.

We may now prove the following analogue of Theorem 1.1: [76]$\|_{77}$

Theorem 1. *Up to T-isomorphism, all and only the sub-icdl's w/t-f of M_0 are T-simple icdl's w/t-f.*

It may be verified that M_0 is T-simple by much the same sort of computational argument that showed D simple. Obviously it suffices to assume that two distinct elements of F_{+0} are T-congruent and then to show under this assumption that all the elements of F_{+0} are T-congruent. Let us first remark that F_{+0}, considered just as a lattice (forgetting N), is isomorphic to D, so those arguments we made concerning the simplicity of D which did not use N may be applied here. Thus $+0 \equiv +3$ may be handled like $0 \equiv 1$, and $+1 \equiv +2$ may be handled like $p \equiv q$. Considerations of symmetry convince us that out of the remaining cases we may restrict our attention to the case where $+0 \equiv +1$ and the case where $+2 \equiv +3$. If $+0 \equiv +1$, then

$-0 \equiv -1$ and hence $+0 \vee -0 \equiv +1 \vee -1$, which is just $+3 \equiv +1$. But from this and our assumption that $+0 \equiv +1$, we get $+0 \equiv +3$, which has already been treated. If we assume that $+2 \equiv +3$, then $+2 \wedge +1 \equiv +3 \wedge +1$, which is just $+0 \equiv +1$, which is just the previous case treated. So M_0 is T-simple.

It is further easily verified that all the sub-icdl's w/t-f of M_0 are T-simple by arguments of the same sort as the above.

The proof of the "only" part of the theorem is like the proof of the "only" part of Theorem 1, except that it uses Theorem II:1.3 in place of Theorem III:1.3. [77]‖[78] Suppose that $(A, \leq, \mathrm{N}, T)$ is a T-simple icdl w/t-f. We may assume that there are at least two distinct elements in T (since otherwise we have only the two-element chain, which is obviously T-isomorphic to a sub-icdl w/t-f of M_0, namely, $\{-3, +3\}$). Then we know by Theorem II:1.3 that there exists a T-homomorphism h of A into M_0 that carries the two distinct elements into two distinct elements. This means that the T-congruence determined by h is not the t-f congruence. But if it is the identity congruence, then h must be one-to-one, in which case A is T-isomorphic to a sub-icdl w/t-f of M_0, which completes the proof.

3.   Normality

The project in this section is to find some reasonably interesting property that D and M_0 have in common, and that somehow expresses the fact that they are large enough (complex enough) to serve as exemplars (via their homomorphism theorems) of De Morgan lattices and icdl's, respectively. This property of "normality" should hence be intimately connected with the respective homomorphism theorems. We add as an afterthought that it would be pleasant if the two element Boolean algebra **2** should also turn out to have this property. These multifarious requirements suggest that the appropriate property, in order to express the "sameness in difference" of D, M_0, and **2**, must be somehow relativized to the relevant [78]‖[79] classes of De Morgan lattices.

We first observe that in a particular class K of De Morgan lattices there may be lattices that have a prime filter that contains for some element, both it and its De Morgan complement; that have a prime filter that contains for some element, neither it nor its complement; or that have a prime filter that contains for some element, it but not its complement.[1] Let us call each of these abstract possibilities a *complementation situation*, and we shall say that a given complementation situation holds with respect to a certain

---

[1]One might first think that there might be yet another situation, namely, that a prime filter contain the complement of an element but not the element itself. But since De Morgan complementation is of period two, this is quickly seen to be indistinguishable from the last situation above.

prime filter P if there is an element and its complement that are included or excluded from P in the appropriate fashion. We shall then say that a given complementation situation exists in the class K if there is some De Morgan lattice $A \in K$ that has some prime filter P with respect to which the complementation situation holds.

Let us then call a De Morgan lattice $A \in K$ *normal in the class K* iff it has a prime filter P with respect to which all of the complementation situations existing in the class hold. When there is no danger of ambiguity, we shall call a De Morgan lattice *normal* (simpliciter) if it is $^{79}\|_{80}$ normal in the class of De Morgan lattices.

Illustrations should clarify these concepts. In the class of De Morgan lattices in general, all of the complementation situations exist, as may be verified by considering D and F_p: F_p includes both p and Np, excludes both q and Nq, and includes 1 but excludes N1. Since all of these complementation situations hold with respect to F_p, D is normal. Similarly, all of the complementation situations exist in icdl's, as may be verified by considering M_0 and F_{-1}, which at the same time shows M_0 normal in icdl's. Since an icdl is thus normal in icdl's iff it is normal in De Morgan lattices, we do not need in the sequel to consider normality in icdl's as distinct from normality in De Morgan lattices (plain normality).

But in Boolean algebras, not all of the complementation situations exist, since every prime filter in a Boolean algebra is maximal and contains for every element, exactly one of it and its complement (cf. Chapter I, Section 3). So normality in Boolean algebras must be distinguished from normality in De Morgan lattices (plain normality). Since prime filters exist in all but the one element Boolean algebra **1**, every Boolean algebra except **1** is Boolean normal, and in particular **2** is Boolean normal. This means that the concept of normality is of no permanent interest in the study of Boolean algebras.

Note that besides Boolean algebras, we have other examples of De Morgan lattices that are not normal. Thus $^{80}\|_{81}$ no chain is normal. The question thus arises for De Morgan lattices and icdl's as to which of them are normal. We can see immediately from Theorems III:1.1 and III:1.2 regarding the mutual determination of prime filters and homomorphisms into D, the following:

Theorem 1. *A necessary and sufficient condition for a De Morgan lattice to be normal is that it have a homomorphism onto D.*

Although an icdl is a De Morgan lattice and normality for icdl's is normality, it is interesting to note that we get a special analogous theorem for icdl's:

Theorem 2. *A necessary and sufficient condition for an icdl to be normal* *is that it have a homomorphism onto M_0.*

The sufficiency follows immediately from Theorem II:1.2. The necessity follows from Theorem II:1.1 *via* the following observations: If an icdl is normal, then it must have a prime filter P and elements a and b such that $a, Na \in P$ and $b, Nb \in \overline{P}$. Now consider some truth filter T of A. T contains exactly one of a and Na, and exactly one of b and Nb, so by the conditions that determine the homomorphism h of Theorem II:1.1, h carries elements of A into $+1$, -1, $+2$, and -2. But since these four "target" elements generate M_0 (in particular, every element of M_0 is a meet or join of these elements), h is easily seen to be onto M_0. [81]‖[82]

Note how these last two theorems square with the fact that a necessary and sufficient condition for a Boolean algebra to be normal in Boolean algebras is that it have a homomorphism onto **2**, which fact rests upon the mutual determination of prime filters and 2-valued homomorphisms. Of course, as we have observed, it so happens that every Boolean algebra but **1** is normal in Boolean algebras, but the parallel remains.

We next prove that a rather simple condition on the elements of a De Morgan lattice is equivalent to normality:

Theorem 3. *A necessary and sufficient condition for a De Morgan lattice to be normal is that it have elements a and b such that $a \wedge Na \not\leq b \vee Nb$.*[2]

Necessity follows by observing that if a De Morgan lattice is normal, then it has a prime filter P and elements a and b such that $a, Na \in P$ and $b, Nb \in \overline{P}$. But then $a \wedge Na \in P$ and $b \vee Nb \in \overline{P}$, and so $a \wedge Na \not\leq b \vee Nb$ since otherwise $b \vee Nb \in P$.

Sufficiency follows by observing that if $a \wedge Na \not\leq b \vee Nb$, then by Stone 1937 there exists a prime filter P such that $a \wedge Na \in P$ and $b \vee Nb \in \overline{P}$. But then $N(a \wedge Na) = Na \vee a \in P$, and $N(b \vee Nb) = Nb \wedge b \in \overline{P}$. Further, $a \vee b \in P$ and $N(a \vee b) = Na \wedge Nb \in \overline{P}$, [82]‖[83] which means that all of the complementation situations hold with respect to P.

It is interesting to note that since $p \,\&\, \overline{p} \to q \vee \overline{q}$ is not a theorem of E or R (cf. Chapter I, Section 1), then in the Lindenbaum algebras of E and R, which are icdl's,[3] $|p| \wedge N|p| \not\leq |q| \vee N|q|$, which gives us

Theorem 4. *The Lindenbaum algebras of E and R are normal.*

This means that the normal De Morgan lattices (icdl's in particular) are of special importance in the study of E and R. So our approving epithet

---

[2]I am indebted to Peter Woodruff for suggesting this result by his first showing that a sufficient condition for an icdl to have a homomorphism onto M_0 is that it contain elements a and b like those above.

[3]Cf. Chapter VII for definition and proof.

"normal" has logical motivations, as well as the algebraic motivations that we have already discussed.

Given this, it is curious to note that Kalman 1958 defined a normal De Morgan lattice as one such that for all elements a and b, $a \wedge Na \leq b \vee Nb$. In short, what is normal for Kalman is abnormal for us, and vice versa. We find this humorous since it was only after we had formulated our notion of normality that we came across Kalman's. Of course, there is nothing wrong with Kalman's notion, just as there is nothing wrong with ours. It is just that due to independent motivations we find different things to approve of in De Morgan lattices (Kalman was seeking a common abstraction of Boolean algebras and l-groups, both of which are normal in his sense and abnormal in ours). Despite the [83]$\|_{84}$ risk of confusion, we retain our notion of normality as a moral lesson that normality depends, in mathematics at least, upon your point of view.

It is worth remarking that one can obtain a whole list of conditions equivalent to normality in our sense by simply negating Kalman's list of conditions equivalent to normality in his sense. [84]$\|_{85}$

V. EMBEDDING ICDL'S IN TERTIARY PRODUCTS OF BOOLEAN ALGEBRAS

If one looks at just the underlying lattice structure of M_0 (forgetting, for the moment, the particular intensional complementation that has been defined upon it), it turns out to be a Boolean algebra. Indeed it turns out to be a direct product of the two element Boolean algebra **2** taken three times, as is shown in the following diagram:

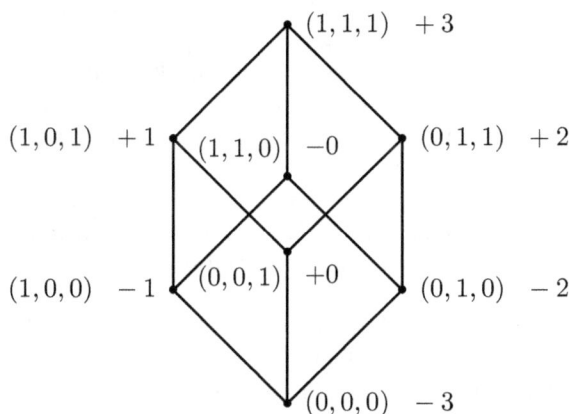

One naturally wonders whether an intrinsic operation that coincides with the defined intensional $^{85}\|_{86}$ complementation of M_0 can be constructed upon this product utilizing in some way the Boolean complementation on **2**. If we define $N(a_1, a_2, a_3) = (\overline{a_2}, \overline{a_1}, \overline{a_3})$, where each $a_i \in \mathbf{2}$ and $\overline{a_i}$ is the Boolean complement of a_i in **2**, it is easily seen that this is just the intensional complementation defined upon M_0.

One next wonders whether this is just an *ad hoc* trick that happens to work for M_0 or whether, and how, it generalizes to products of M_0 (and hence through the embedding Theorem II:2.1 to arbitrary icdl's).

Let us then attempt to generalize the representation of M_0 as a product of Boolean algebras (with the funny sort of "flip" complementation). We define a *tertiary product* (*3-product*) of a Boolean algebra B as a structure $(B^3, \leq, N)$, where the elements of B^3 are the triples (a_1, a_2, a_3) where each $a_i \in B$, where $(a_1, a_2, a_3) \leq (b_1, b_2, b_3)$ iff $a_i \leq b_i$ in B, and where $N(a_1, a_2, a_3) = (\overline{a_2}, \overline{a_1}, \overline{a_3})$ where $\overline{a_i}$ is the Boolean complement of a_i in B.

In short, a 3-product of B is just the ordinary direct product of B taken three times, except for complementation, which, instead of being defined simply component-wise, has a "flip" thrown in on the first two components as well. A 3-product is easily seen to be an icdl.

We shall show that every product of M_0 is embeddable in a 3-product of Boolean algebras. We shall show this by showing in particular that every icdl is embeddable in a 3-product of a direct product of **2**. Let [86]$\|$[87] M^c be a product of M. We shall embed M^c in a 3-product of a direct product of $\{\mathbf{2}_x\}_{x<c}$.

Let us first define functions h_1, h_2, h_3 which take an element $a_x \in M_{0_x}$ into $\mathbf{2}_x$. Let $h_1(a_x) = 1_x$ if $a_x \in F_{-1_x}$, and otherwise $h_1(a_x) = 0_x$. Let $h_2(a_x) = 1_x$ if $a_x \in F_{-2_x}$, and otherwise $h_2(a_x) = 0_x$. And let $h_3(a_x) = 1_x$ if $a_x \in F_{+0_x}$, and otherwise $h_3(a_x) = 0_x$. Then let us define for $\{a_x\}_{x<c}$,
$$h(\{a_x\}_{x<c}) = (\{h_1(a_x)\}_{x<c}, \{h_2(a_x)\}_{x<c}, \{h_3(a_x)\}_{x<c}).$$

The one-to-oneness of h follows easily from an observation made in the proof of Theorem II:2.1, namely, that in $M_0 \in \varepsilon F_{-1} \cap \varepsilon F_{-2} \cap \varepsilon F_{+0}$ is always a unit set, for every choice function ε that, remember, chooses for every F_i either F_i or $\overline{F_i}$. Thus if $h(\{a_x\}_{x<c}) = h(\{b_x\}_{x<c})$, then for every $x < c$, $h_1(a_x) = h_1(b_x)$, $h_2(a_x) = h_2(b_x)$, and $h_3(a_x) = h_3(b_x)$, i.e., for every $x < c$, $a_x \in F_{-1_x}$ iff $b_x \in F_{-1_x}$, $a_x \in F_{-2_x}$ iff $b_x \in F_{-2_x}$, and $a_x \in F_{+0_x}$ iff $b_x \in F_{+0_x}$, which means in view of our initial observation that for every $x < c$, $a_x = b_x$.

That h preserves meets, i.e., that $h(\{a_x\}_{x<c} \wedge \{b_x\}_{x<c}) = h(\{a_x\}_{x<c}) \wedge h(\{b_x\}_{x<c})$ quickly reduces to the fact that for $i = 1, 2, 3$, $h_i(a_x \wedge b_x) = h_i(a_x) \wedge h_i(b_x)$, which fact is a special case of the well known fact that prime filters determine homomorphisms into **2**.

The h preserves N, i.e., that $h(\mathrm{N}\{a_x\}_{x<c}) = \mathrm{N}h(\{a_x\}_{x<c})$ reduces to $h_1(\mathrm{N}a_x) = h_2(a_x)$, $h_2(\mathrm{N}a_x) = h_1(a_x)$, and $h_3(\mathrm{N}a_x) = h_3(a_x)$. The first identity is a consequence of the [87]$\|$[88] fact that in M_0 $\mathrm{N}(F_{-1}) = \overline{F_{-2}}$. Thus $\mathrm{N}a_x \in F_{-1_x}$ iff $\mathrm{NN}a_x = a_x \in \mathrm{N}(F_{-1_x}) = \overline{F_{-2_x}}$. Similarly the second identity follows from the fact that $\mathrm{N}(F_{-2}) = \overline{F_{-1}}$, and the third follows from the fact that $\mathrm{N}(F_{+0}) = \overline{F_{+0}}$. Thus we have proven

Theorem 1. *Every product of M_0, M^c, is embeddable in a 3-product of a direct product of **2**, hence in a 3-product of a Boolean algebra.*

Since we know by Theorem II:2.1 that any icdl is embeddable in a product of M_0, we get as a corollary

Theorem 2. *Every icdl is embeddable in a 3-product of a direct product of **2**, hence in a 3-product of a Boolean algebra.*

It is interesting to note that with the exception of M_0 itself, no product of M_0, M^c, is isomorphic to a 3-product of a direct product of **2** under the mapping h defined in the proof of Theorem 1. That then h fails to be onto is readily seen once it is observed that M^c was defined so that it cannot have an element $\{a_x\}_{x<c}$ which has components, say a_{x_1} and a_{x_2}, such that $a_{x_1} \in F_{+0_{x_1}}$ and $a_{x_2} \in F_{+0_{x_2}}$. Thus no element of the 3-product is in the image of h unless its third component is an indexed set of all 0's or all 1's. This reflects the decision in Belnap's definition of the product M^c to not allow an element that has some "true" components (from $\overline{F_{+0}}$) and some "false" components (from F_{+0}). However, it is obvious that any M^c is isomorphic under h to a special sub-3-product of the direct product of $\{\mathbf{2}_x\}_{x<c}$, namely, $^{88}\|_{89}$ that sub-3-product that consists of all elements of the form $(\{a_{1_x}\}_{x<c}, \{a_{2_x}\}_{x<c}, \{a_{3_x}\}_{x<c})$, where $a_{1_x}, a_{2_x} \in \mathbf{2}_x$ and where every $a_{3_x} = 0_x$ or every $a_{3_x} = 1_x$. Let us call such a special sub-3-product a *special sub-3-product*. To see that h is onto a special sub-3-product is simply to see that there are elements in F_{+0} that are in both F_{-1} and F_{-2} ($+3$), in F_{-1} and $\overline{F_{-2}}$ ($+1$), in $\overline{F_{-1}}$ and F_{-2} ($+2$), and in $\overline{F_{-1}}$ and $\overline{F_{-2}}$ ($+0$), and that similarly there are elements in $\overline{F_{+0}}$ for every possibility. Thus we get

Theorem 3. *Every product of M_0, M^c, is isomorphic to a special sub-3-product of a direct product of* $\mathbf{2}$.

So products of M_0 and special sub-3-products of direct products of $\mathbf{2}$ may be abstractly identified, which may have some utility with regards to the semantic applications of products of M_0 to intensional logics.

Leaving this curiosity, we go on to observe

Theorem 4. *Every icdl is embeddable in a 3-product of a field of sets.*

We can see this at once from Theorem 2, which tells us that every icdl is embeddable in a 3-product of a direct product of $\mathbf{2}$, together with the familiar fact that every direct product of $\mathbf{2}$ is isomorphic with a field of sets. All we have to do is to replace the direct product of $\mathbf{2}$ in Theorem 2 with its isomorphic field of sets. It is worth remarking that this move from Theorem 2 to Theorem 4 does not require the axiom of choice or any of its equivalents, and is thus *effective* in the $^{89}\|_{90}$ sense of Sikorski 1964. This is because the theorem that a direct product of $\mathbf{2}$ is isomorphic to a field of sets is effectively provable (cf. the proof in Birkhoff 1948, p. 140). Similarly, our move from Theorem II:2.1 to Theorem 2 was effective, as may be checked by reviewing its proof. We now show that Theorem 4 effectively implies Theorem II:2.1, thus completing the daisy chain of effective implications to give us

Theorem 5. *Theorems II:2.1, 2, and 4 are effectively equivalent, i.e., anyone may be proven from the assumption of any other without using some equivalent of the axiom of choice.*

Assume that an icdl A is embedded in a sub-3-product of a field F of sets all of which are subsets of a set X. Let the embedding isomorphism be f. If the cardinality of X is c, we shall embed A in M^c. Let us first remember that f takes an element a into a triple of sets (X_1, X_2, X_3). Let us define new functions f_1, f_2, and f_3 so that $f_1(a) = X_1$, $f_2(a) = X_2$, and $f_3(a) = X_3$. We then define a mapping h of A into M^c so that $h(a) = \{h_x(a)\}_{x \in X}$, where h_x is a mapping into M_0 defined as satisfying the following conditions:

i) $h_x(a) \in F_{-1}$ iff $x \in f_1(a)$,

ii) $h_x(a) \in F_{-2}$ iff $x \in f_2(a)$,

iii) $h_x(a) \in F_{+0}$ iff $x \in f_3(a)$.

That these conditions do indeed determine (single-valued) functions h_x (and hence a (single-valued) function h) may $^{90}\|_{91}$ be shown by the same argument used in the proof of Theorem II:1.1. The one-to-oneness of h follows trivially from that of f. And h may be shown to be a homomorphism by showing each h_x a homomorphism (since operations in M^c are defined component-wise). The strategy of proof is similar to that of Theorem II:1.2. We show that $h_x(a \wedge b) \in F_i$ iff $h_x(a) \wedge h_x(b) \in F_i$ $(i = -1, -2, +0)$, which suffices to show $h_x(a \wedge b) = h_x(a) \wedge h_x(b)$ in virtue of the way h_x was defined. We treat the case $i = -1$, which is representative of the other two cases: $h_x(a \wedge b) \in F_{-1}$, iff $x \in f_1(a \wedge b)$, iff $x \in f_1(a) \cap f_1(b)$, iff $h_x(a), h_x(b) \in F_{-1}$, iff $h_x(a) \wedge h_x(b) \in F_{-1}$. Note that the only special property of the set F_{-1} that we used was that it is a filter, so the case is indeed representative. We next show that $h_x(\mathrm{N}a) \in F_i$ iff $\mathrm{N}(h_x a) \in F_i$ $(i = -1, -2, +0)$. We treat the case $i = -1$, using the fact that $F_{-1} = \mathrm{N}(\overline{F_{-2}})$: $h_x(\mathrm{N}a) \in F_{-1}$, iff $x \in f_1(\mathrm{N}a)$, iff $x \in \overline{f_2(a)}$, iff $h_x(a) \in \overline{F_{-2}}$ iff $\mathrm{N}h_x(a) \in F_{-1}$. The cases $i = -2$ and $i = +0$ follow analogously using respectively the facts $F_{-2} = \mathrm{N}(\overline{F_{-1}})$ and $F_{+0} = \mathrm{N}(\overline{F_{+0}})$, which completes the proof.

It is perhaps worthwhile to point out that although Theorems II:2.1, 2, and 4 are effectively equivalent, that we have no effective proof of them. The proof of Theorem II:2.1, which is our basis for the other two theorems, rests upon Stone's 1937 theorem concerning the separation of elements in a distributive lattice by prime filters, the proof of which requires Zorn's lemma, an equivalent of the axiom of choice. The situation seems parallel to the $^{91}\|_{92}$ situation in Boolean algebra, where a number of fundamental theorems (e.g., the representation and embedding theorems) are effectively equivalent, but no proof of them exists that does not use the axiom of choice (for an excellent discussion of this cf. Sikorski 1964). The situation may be made to seem more parallel by showing that there are "direct" proofs of Theorems 2 and 4 that do not rely (in appearance, at least) upon the proof of Theorem II:2.1. This is comparable to the well-known "independent" but equally non-effective proofs of the fundamental theorems of Boolean algebra.

Thus Theorem 4 may be proven by defining on an arbitrary icdl A, with prime filters P and truth filters T, a mapping: $h(a) = (\{P\colon a \in P\}, \{P\colon \mathrm{N}a \in \overline{P}\}, \{T\colon a \in T\})$. Upon observing that every truth filter is a prime filter, it is not difficult to prove that h is an isomorphism into the 3-product of the field of all subsets of the set of prime filters of A. But the one-to-oneness of h depends, of course, on Stone's prime filter separation theorem.

And we may prove Theorem 2 for an arbitrary icdl A by first picking a certain truth-filter T (so that we are in effect considering the icdl w/t-f (A, T)). We next consider the indexed set $\{P_x\}_{x \in X}$ of all prime filters of A. If X has cardinality c, we map A into a 3-product of a direct product of $\{2_x\}_{x<c}$ by the following mapping h: $h(a) = (\{h_{1_x}(a)\}_{x \in X}, \{h_{2_x}(a)\}_{x \in X}, \{h_{3_x}(a)\}_{x \in X})$, where $h_{1_x}(a) = 1$ if $a \in P_x$, and otherwise 0; where $h_{2_x}(a) = 1$ if $^{92}\|_{93}$ $\mathrm{N}a \in \overline{P_x}$, and otherwise 0; and where $h_{3_x} = 1$ if $a \in T$, and otherwise 0. The mapping h may be readily verified to be a homomorphism, but again the one-to-oneness of h depends upon Stone's prime filter theorem. $^{93}\|_{94}$

VI. Embedding De Morgan Lattices in Binary Products of Boolean Algebras

As usual, whatever sort of treatment can be given to icdl's can be done more simply for De Morgan lattices in general. Thus if one looks at just the underlying lattice structure of D, it turns out to be a direct product of the two element Boolean algebra **2** taken two times, as is shown in the following diagram:

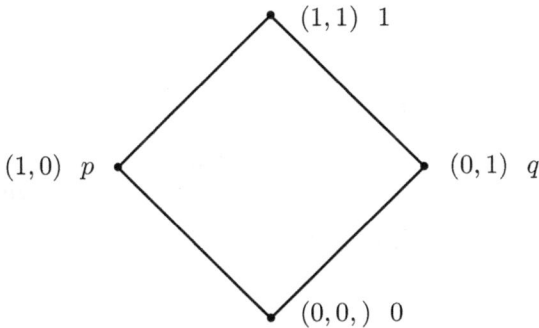

$$
\begin{array}{ccc}
 & (1,1)\ \ 1 & \\
(1,0)\ \ p & & (0,1)\ \ q \\
 & (0,0,)\ \ 0 &
\end{array}
$$

If we define $N(a_1, a_2) = (\overline{a_2}, \overline{a_1})$, where each $a_i \in \mathbf{2}$ and $\overline{a_i}$ is the Boolean complement of a_i in **2**, it is easily seen that this is just the original De Morgan complementation defined upon D. This suggests that the appropriate structure in which to embed De Morgan lattices (in analogy with our embedding of icdl's in 3-products of Boolean algebras) [94][95] is a *binary product (2-product)* of Boolean algebras, defined as follows: A 2-product of a Boolean algebra B is a triple $(B^2, \leq, N)$, where the elements of B^2 are the pairs (a_1, a_2) where each $a_i \in B$, where $(a_1, a_2) \leq (b_1, b_2)$ iff $a_i \leq b_i$ in B, and where $N(a_1, a_2) = (\overline{a_2}, \overline{a_1})$ where $\overline{a_i}$ is the Boolean complement of a_i in B. Note that this is just like our definition of 3-product in the last chapter except that the third component has been omitted. A 2-product is easily seen to be a De Morgan lattice.

We may now prove theorems analogous to those of the last chapter for icdl's and 3-products. Thus we get

Theorem 1. *Every product of D, D^c, is isomorphic to a 2-product of a direct product of **2**, hence to a 2-product of a Boolean algebra.*

The proof of this theorem is strictly analogous to the proof of Theorem V:1. We embed D^c in a 2-product of the direct product of $\{2_x\}_{x<c}$ by defining mappings h_1 and h_2 which take an element $a_x \in D_x$ into 2_x. Let $h_1(a_x) = 1_x$ iff $a_x \in F_p$, and let $h_2(a_x) = 1_x$ iff $a_x \in F_q$. Then define for an element of D^c, $\{a_x\}_{x<c}$, $h(\{a_x\}_{x<c}) = (\{h_1(a_x)\}_{x<c}, \{h_2(a_x)\}_{x<c})$. To show that h is an isomorphism is now just copy work from Theorem V:1, which we leave unwritten. But unlike the embedding of Theorem V:1, which was "almost onto," this embedding h is onto. There is hence no need to talk of "special" sub-2-products like the special sub-3-products of Theorem V:3. To see that h is onto is simply to see $^{95}\|_{96}$ that there are elements in D that are in both F_p and F_q (1), in F_p and $\overline{F_q}$ (p), in $\overline{F_p}$ and F_q (q), and in $\overline{F_p}$ and $\overline{F_q}$ (0).

Now since we know by Theorem III:2.1 that any De Morgan lattice is embeddable in a product of D, we get

Theorem 2. *Every De Morgan lattice is embeddable in a 2-product of a direct product of* **2**, *hence in a 2-product of a Boolean algebra.*

From Theorem 2 and the well-known fact that every direct product of **2** is isomorphic to a field of sets (cf. proof of Theorem V:4), we get

Theorem 3. *Every De Morgan lattice is embeddable in a 2-product of a field of sets.*

We remark that in the process of proving Theorems 2 and 3 we have shown that Theorem III:2.1 effectively implies Theorem 2, and that Theorem 2 effectively implies Theorem 3 (cf. remarks preceding Theorem V:5 for elucidation of this remark). So by now showing that Theorem 3 effectively implies Theorem III:2.1, we prove the following analogue to Theorem V:5:

Theorem 4. *Theorems III:2.1, 2, and 3 are effectively equivalent.*

We assume that a De Morgan lattice A is embedded in a sub-2-product of a field F of sets all of which are subsets of a set X, in accord with Theorem 3. Let the embedding isomorphism be f. If the cardinality of X is c, we shall embed A in D^c, in accord with Theorem III:2.1. $^{96}\|_{97}$ Let us suppose that f takes an element a into a pair of sets (X_1, X_2), and then let us define new functions f_1 and f_2 so that $f_1(a) = X_1$ and $f_2(a) = X_2$. We then define a mapping h of A into D^c so that $h(a) = \{h_x(a)\}_{x \in X}$ where h_x is a mapping into D defined as satisfying the following:

i) $h_x(a) \in F_p$ iff $x \in f_1(a)$,

ii) $h_x(a) \in F_q$ iff $x \in f_2(a)$.

From this point on the proof proceeds just like that following Theorem V:5.

The remarks that we should like to make about the effectivity situation are exactly parallel to the remarks that we made after the proof of Theorem V:5. The gist of them is that although Theorems III:2.1, 2, and 3 were shown equivalent without using the axiom of choice, all known proofs of any of them seem to require the axiom of choice. Thus we covertly used Zorn's lemma in the proof of Theorem III:2.1, basing it as we did on Stone's prime filter separation theorem. And we can give "independent" proofs of Theorems 2 and 3, but they all rely for the one-to-oneness of the embeddings upon Stone's theorem. Thus Theorem 3 may be proven by defining on an arbitrary De Morgan lattice A, with prime filters P, a mapping: $h(a) = (\{P: a \in P\}, \{P: Na \in \overline{P}\})$. And we may prove Theorem 2 by first considering the indexed set $\{P_x\}_{x<c}$ of all prime filters of a De Morgan lattice A. We map A into a 2-product of a direct product of $\{\mathbf{2}_x\}_{x<c}$ by the following mapping h: $h(a) = (\{h_{1_x}(a)\}_{x<c}, \{h_{2_x}(a)\}_{x<c})$, where $^{97}\|_{98}$ $h_{1_x}(a) = 1$ if $a \in P_x$, and otherwise 0; and where $h_{2_x}(a) = 1$ if $Na \in \overline{P_x}$, and otherwise 0.

It is interesting to compare Theorem 2 with a representation theorem for De Morgan lattices due to Białynicki-Birula and Rasiowa 1957. Before stating this theorem, we first recall some appropriate terminology. A *permutation* g on a set X is a one-to-one mapping of X onto itself. If the permutation g is of period two, i.e., if $g(g(x)) = x$, then g is called an *involution*.[1] Białynicki-Birula and Rasiowa consider a set X together with an involution g, and define for every $Y \subseteq X$, an operation $N_g(Y) = \overline{g(Y)}$. They then define a *quasi-field of sets* as a ring of subsets of such a set X with involution g that is closed under the operation N_g. They observe that every quasi-field of sets is a De Morgan lattice, and they further show, using Stone's 1937 Prime Filter Theorem, that every De Morgan lattice is isomorphic to a quasi-field of sets. Their representation theorem may be looked upon as a generalization of Stone's 1936 representation theorem for Boolean algebras, since a detailed statement of their theorem would include the fact that when the De Morgan lattice is a Boolean algebra, then g is simply the identity $^{98}\|_{99}$ function and the quasi-field of sets is a (genuine) field of sets. We shall eventually demonstrate that our Theorem 2 is effectively equivalent to their representation theorem, but for so doing, we find it convenient to first develop another effective equivalent of their theorem,

---

[1] Note that this sense of *involution* differs from the sense of *involution* as a dual *lattice* automorphism of period two. The latter is a special lattice theoretical notion, whereas the former is a general set theoretical notion. However, Białynicki-Birula and Rasiowa's representation theorem reveals a surprising interconnection between the two notions.

which has independent interest as well. Just as Białynicki-Birula and Rasiowa's representation theorem may be looked upon as a generalization of Stone's representation theorem for Boolean algebras, so this effective equivalent may be looked upon as a generalization of Stone's embedding theorem for Boolean algebras, which we recall states that every Boolean algebra is embeddable in a direct product of **2**.

We accordingly define a *direct quasi-product of a Boolean algebra B* as follows:[2] Consider an indexed set of replicas of B, $\{B_x\}_{x \in X}$, together with a permutation g on the indexing set X. Elements of the direct quasi-product are members of the Cartesian product $\times_{x \in X} B_x$, and meet and join are defined upon these elements component-wise, as usual. But where $\{a_x\}_{x \in X}$ is an element, we define $N_g(\{a_x\}_{x \in X}) = \{\overline{a_{g(x)}}\}_{x \in X}$. A direct quasi-product of a Boolean algebra is just like a (genuine) direct product of that Boolean algebra, except for the funny kind of complementation, where one first permutes, then applies $\|$ the ordinary Boolean complement. It is easy to see that any quasi-direct product of a Boolean algebra is a De Morgan lattice. We now prove

Theorem 5. *Every De Morgan lattice is embeddable in a direct quasi-product of* **2**,

by showing at the same time that this theorem is effectively equivalent to Białynicki-Birula and Rasiowa's representation theorem. We show this in turn by giving an effective proof of

Theorem 6. *Every direct quasi-product of* **2** *is isomorphic to a quasi-field of* all *subsets of some set, and conversely, every quasi-field of* all *subsets of some set is isomorphic to some direct quasi-product of* **2**, *which means that every quasi-field of subsets of some set is embeddable in some direct quasi-product of* **2**.

For the proof, we first recall that every element of a Cartesian product $\times_{x \in X} \mathbf{2}_x$ may be looked at as a characteristic function defined upon the set of indices X.[3] This is to say that each element $\{a_x\}_{x \in X}$ determines its characteristic subset Y of X so that $x \in Y$ iff $a_x = 1$, and conversely that every subset Y of X determines its characteristic function $\{a_x\}_{x \in X}$ so that $a_x = 1$ iff $x \in Y$. The mapping which sends every element of a direct

---

[2]Note in what follows that the definition is made only for the case where B is a single Boolean algebra, because only then are we guaranteed closure under the defined N. The notion of a direct quasi-product of a set of *distinct* Boolean algebras does not make the sense that the notion of a (genuine) direct product of the same does.

[3]This observation forms the basis for Stone's 1936 proof that every (genuine) direct product of **2** is isomorphic to a (genuine) field of all subsets of some set, and our proof is based upon his.

quasi-product $^{100}\|_{101}$ on the Cartesian product $\times_{x \in X} \mathbf{2}_x$ with permutation g into the characteristic set determined by the element is easily seen to be an isomorphism onto the quasi-field of all subsets of X with permutation g. And conversely, the mapping which sends every subset of a set X into its characteristic function $\{a_x\}_{x \in X}$ is easily seen to be an isomorphism of the quasi-field of all subsets of X with permutation g onto a direct quasi-product based on $\times_{x \in X} \mathbf{2}_x$ with permutation g, which completes the proof.

We next show that Theorem 5 is effectively equivalent to Theorem 3. Since we have just shown that Białynicki-Birula and Rasiowa's representation theorem is effectively equivalent to Theorem 5, and since Theorem 4 states that Theorem 3 is effectively equivalent to Theorem 2, this will give us the desired result that Białynicki-Birula and Rasiowa's representation theorem is effectively equivalent to our representation Theorem 2. We show that Theorem 5 is effectively equivalent to Theorem 2 by giving an effective proof of the following,

Theorem 7. *Every 2-product of a direct product of* $\mathbf{2}$ *is isomorphic to a direct quasi-product of* $\mathbf{2}$, *and conversely, every direct quasi-product of* $\mathbf{2}$ *is embeddable in a 2-product of a direct product of* $\mathbf{2}$.

The first part of the theorem is almost immediate, for roughly speaking, every 2-product of a direct product of $\mathbf{2}$ *is* a direct quasi-product of $\mathbf{2}$. All that is required is some resubscripting to make this precise. For consider $^{101}\|_{102}$ an element $(\{a_{1_x}\}_{x \in X}, \{a_{2_x}\}_{x \in X})$ of a 2-product of a direct product of $\mathbf{2}$. It may be looked at as a pair of functions from the indexing set X into $\mathbf{2}$. If we consider another (disjoint) indexing set X^* that is in a one-to-one correspondence with X (each element $x \in X$ corresponding to an element $x^* \in X^*$), we can map each element $(\{a_{1_x}\}_{x \in X}, \{a_{2_x}\}_{x \in X})$ into $(\{a_{1_x}\}_{x \in X}, \{a_{2_{x^*}}\}_{x^* \in X^*})$. This is a mapping of the elements of a 2-product of a direct product of $\mathbf{2}$ into the Cartesian product $\times_{y \in X \cup X^*} \mathbf{2}_y$. When this Cartesian product is made into a direct quasi-product by defining a permutation g on the set of indices $X \cup X^*$ so that for $x \in X$, $g(x) = x^*$, and for $x^* \in X^*$, $g(x^*) = x$, then this mapping may be easily seen to be an isomorphism of the 2-product of the direct product of $\mathbf{2}$ onto a direct quasi-product of $\mathbf{2}$.

To show that every direct quasi-product of $\mathbf{2}$ is embeddable in a 2-product of a direct product of $\mathbf{2}$, we consider such a direct quasi-product defined upon $\times_{x \in X} \mathbf{2}_x$ with permutation g, and we map each element $\{a_x\}_{x \in X}$ into $(\{a_x\}_{x \in X}, \{a_{g(x)}\}_{x \in X})$, which is an element of the 2-product of the direct product of $\{\mathbf{2}_x\}_{x \in X}$. This mapping is easily verified to be an isomorphism.

We now have quite a batch of effective equivalents. Referring to Theorem 4, and the text which proceeded it, we simply record these in

Theorem 8. *Theorems III:2.1 (Kalman's 1958 embedding theorem), 2, 3, 5, and Białynicki-Birula and* [102]||[103] *Rasiowa's 1957 representation theorem are all effectively equivalent.* [103]||[104]

VII. COMPLETENESS OF FIRST DEGREE ENTAILMENTS

1. First Degree Entailments

This chapter, as well as the next, will be concerned with recasting into an algebraic framework certain results of Anderson and Belnap, and of Smiley as regards the completeness and consistency of a certain fragment of E and R.[1] As mentioned in Section 4 of Chapter I, such a recasting of logical results into an algebraic framework has two aspects. The first is a substitution of the terminology of the algebraist for the terminology of the logician. The second aspect is a more profound substitution of the proof methods of the algebraist for the proof methods of the logician.

Let us begin by describing the appropriate fragment of E and R. We define a *zero degree formula* (*zdf*) in E or R as a formula that is purely truth functional, i.e., one in which occur only the connectives of negation and disjunction (remembering that conjunction can be defined from these in the usual De Morgan manner). We then define a *first degree entailment* (*fde*) in E or R as a formula of the form $A \to B$, where A and B are zdf's. Now following Belnap 1965, let us define a notion of model which relates [104]||[105] fde's to De Morgan lattices in an appropriate fashion, and from this formulate a criterion of semantic completeness.[2]

2. De Morgan Lattices as Models for First Degree Entailments

Let us say that Q is a *De Morgan lattice model* (henceforth in this chapter, *model*) iff $Q = (A, s)$, where A is a De Morgan lattice, and where s is an *assignment function* for A, i.e., a function such that for each propositional variable A, $s(A) \in A$.

Given a model $Q = (A, s)$, we define a (zero degree) *valuation* determined by Q as a function v_Q defined over all zdf's and having values in A as follows: for all zdf's A,

if A is a propositional variable, then $v_Q(A) = s(A)$;

if A has the form $\overline{B}$, then $v_Q(A) = Nv_Q(B)$;

---

[1] Such a recasting was pioneered in Belnap 1965. Cf. Section 5 of Chapter I.

[2] Belnap's definition actually related a slightly larger class of formulas (*first degree formulas*) only to icdl's, but the spirit of our definition is the same.

if A has the form $B \vee C$, then $v_Q(A) = v_Q(B) \vee v_Q(C)$.

Note that these conditions determine that

if A has the form $B \ \& \ C$, then $v_Q(A) = v_Q(B) \wedge v_Q(C)$.

Let us then say that a fde $A \to B$ is *true in a model* $Q = (A, s)$ iff $v_Q(A) \le v_Q(B)$, and otherwise *false in a model*.

We then define a fde as *valid in a De Morgan lattice* A iff it is true in every model $Q = (A, s)$, and [105]$\|_{106}$ otherwise as *falsifiable in the De Morgan lattice* A.

We finally define a fde as *valid in a class of De Morgan lattices* iff it is valid in each member of the class, and otherwise as *falsifiable in the class*. If the class in question is the class of all De Morgan lattices, then we shall say simply that the fde is *valid* or *falsifiable*.

We shall say that the fde fragment of E (or R) is *consistent with respect to a class of De Morgan lattices* iff, if a fde is provable in E (or R), then it is valid in that class. We shall say that the fde fragment of E (or R) is complete with respect to a class of De Morgan lattices iff, if a fde is valid in that class, then it is provable in E (or R). We shall call the fde fragment of E (or R) just *consistent* or *complete* iff the particular class in question is the class of all De Morgan lattices.

We shall eventually relate the provability of a fde with its validity, but first we must take a detour through the Lindenbaum algebras of E and R.

3.   The Lindenbaum Algebras of E and R

Let us define a relation of co-entailment between two statements A and B as holding just when $A \leftrightarrow B$ (an abbreviation for $(A \to B) \ \& \ (B \to A)$) is provable (in E, or in R, depending upon which system we are considering). It is trivially verified that such a relation is a genuine equivalence relation in the sense of Chapter I, Section 2. [106]$\|_{107}$ Further, the relation of co-entailment is a congruence relation preserving negation, disjunction, and entailment, in the sense of Replacement Theorem in Chapter I, Section 1.

All this means that we can introduce an algebra upon the congruence classes of formulas that co-entail one another. Thus if we let $|A|$ be the class of formulas that co-entail A, we may define operations upon these congruence classes as follows. Let

$|A| \vee |B| = |A \vee B|$

$\mathrm{N}|A| = |\overline{A}|$

$|A| \to |B| = |A \to B|$

The point of the Replacement Theorem is to ensure that these operations are well-defined in the sense of not depending upon the particular representatives chosen from the congruence classes. Note incidentally that in the proof of the completeness of the fde fragments we do not need the third of the identities above, since, as will be made clear, we shall focus our attention upon those congruence classes that have as representatives zdf's. But we find it more convenient to make Lindenbaum algebras out of the whole of the systems as we shall have need of them later.

We now have operations defined upon the Lindenbaum algebras. Let us also introduce a relation by defining

$$|A| \leq |B| \text{ iff } A \to B \text{ is provable.} \quad {}^{107}\|_{108}$$

It is now easily verified that the Lindenbaum algebra of E (let us call it $E_{/\leftrightarrow}$) is an icdl. Once we define

$$|A| \wedge |B| = |A \& B| \quad \text{(where } A \& B \text{ is defined as } \overline{\overline{A} \vee \overline{B}}),$$

it follows almost directly from the axioms that $E_{/\leftrightarrow}$ is a distributive lattice and that N is antitone and of period two. It only remains to show that N has no fixed point. But if $N|A| = |A|$ we would have as a theorem $A \leftrightarrow \overline{A}$, which leads in E via A5 (together with double negation and replacement) to both A and $\overline{A}$ as theorems. But E is negation consistent, since E is easily seen to be a subsystem of the classical propositional calculus (mapping the arrow into the horseshoe), which is well-known to be negation consistent. And since E is a subsystem of R, and R is also a subsystem of the classical propositional calculus, we have

Theorem 1. *The Lindenbaum algebras of E and R, $E_{/\leftrightarrow}$ and $R_{/\leftrightarrow}$, respectively, are icdl's.*

4. Completeness of the First Degree Entailment Fragments of E and R

After these preliminaries concerning Lindenbaum algebra, we are able to prove the following completeness theorem.

Theorem 1. *The fde fragments of E and R are complete with respect to the class of icdl's; hence, complete.* ${}^{108}\|_{109}$

We remark that the second clause is a trivial consequence of the first, since an icdl is merely a special kind of De Morgan lattice. We shall prove the first clause by showing that if a fde is not provable in E, then it is falsifiable in the Lindenbaum algebra of E, $E_{/\leftrightarrow}$. An exactly similar proof works for R and $R_{/\leftrightarrow}$.

We begin by defining the *canonical model* C for E as $C = (E_{/\leftrightarrow}, c)$, where c is the *canonical assignment function*, i.e., where A is a propositional variable, $c(A) = |A|$. Let us then define the (zero degree) *canonical valuation* v_C as the (zero degree) valuation determined by C. It is easy to see that where A is a zdf, $v_C(A) = |A|$.

It is now immediate that if a fde $A \to B$ is not provable in E, then $A \to B$ is falsifiable in the icdl $E_{/\leftrightarrow}$ under the canonical valuation v_C, since then $|A| \nleq |B|$ and hence $v_C(A) \nleq v_C(B)$. So we have

Lemma 1. *If a fde is not provable in E (R), then it is falsifiable under the canonical valuation in $E_{/\leftrightarrow}$ ($R_{/\leftrightarrow}$).*

From this lemma, Theorem 1 follows immediately.

5. E, R, M_0 and D

So far we have proven only a rather trivial sort of completeness theorem, the sort that could be proven for any system on which can be defined a well-behaved [109]$\|_{110}$ equivalence relation so as to give rise to an algebra—with or without an attractive name. But Lemma 1, together with Theorems II:1.3, and III:1.3 give rise to the following deeper completeness theorem.

Theorem 1. *A fde is valid in M_0 only if provable in E (and R); a fde is valid in D only if provable in E (and R).*[3]

Again we do the proof for E, the proof for R being exactly similar. We shall prove the contrapositive, first for M_0 and then for D.

On the assumption that the fde $A \to B$ is not provable in E, $|A| \nleq |B|$ in $E_{/\leftrightarrow}$. But since $E_{/\leftrightarrow}$ is an icdl, then by Theorem II:1.3 there exists a homomorphism h of $E_{/\leftrightarrow}$ into M_0 such that $h(|A|) \nleq h(|B|)$. Now define hc as the composition of h and the canonical assignment function c, i.e., if C is a propositional variable, then $hc(C) = h(|C|)$. Obviously hc is an assignment function for M_0. Thus let $Q = (M_0, hc)$. It is easy to see that for all formulas C, $v_Q(C) = h(|C|)$. Hence $v_Q(A) \nleq v_Q(B)$, and $A \to B$ is falsifiable in M_0. The proof to show that $A \to B$ is falsifiable in D is exactly parallel except that it takes [110]$\|_{111}$ starting point in Theorem III:1.3, which shows that there is a homomorphism h of $E_{/\leftrightarrow}$ into D such that $h(|A|) \nleq h(|B|)$. Thus the proof is complete.

---

[3]The first part of this theorem was originally proven in Anderson and Belnap 1962a by normal form methods. The second part was originally proven by Timothy Smiley *via* a modification of the Anderson–Belnap proof. The Smiley result is unpublished, but the author has seen a sketch of the proof in a copy of a letter written by Smiley to Casimir Lewy, dated September 11, 1965. This copy was shown to me by Professor Anderson.

6. An Incomplete Proof of the Completeness of First Degree Entailments with Quantifiers

Machinery for quantification can be added to E and R in much the usual way to obtain the resulting systems EQ and RQ; and upon defining a first degree entailment in EQ or RQ as a formula of the form $A \rightarrow B$ where A and B contain only truth functions and quantifiers (no arrows), appropriate completeness and consistency results with respect to the class of complete icdl's have been obtained by Anderson and Belnap.[4]

It might be expected that we would here include an algebraic proof of the completeness results, but difficulties prevent us from doing so. We shall instead sketch what seems to be a straightforward attempt at such a proof, and remark what remains to be proven.

Let us begin by assuming that EQ and RQ are formulated from a stock of predicate constants and individual variables with the usual formation rules. We must then modify our notion of a model from the propositional $^{111}\|_{112}$ case. Following in spirit Belnap 1965, we define a *model* as a triple $Q = (A, I, s)$, where A is a complete De Morgan lattice, I is a non-empty set of individuals, and s is an *assignment function* for A and I, i.e., if x is an individual variable, $s(x) \in I$, and if F is an n-ary predicate, $s(F)$ is a function from I^n into A.

We may now modify our notion of a *valuation* determined by Q from the propositional case. We simply replace the base clause, which evaluates propositional variables, with the following:

> if A is of the form $Fx_1 \ldots x_n$, $v_Q(A)$ is the value of the function $s(F)$ for the argument $(s(x_1), \ldots, s(x_n))$.

And then we add on to the inductive clauses the following:

> if A has the form $(Ex)B$, $v_Q(A) = \bigvee B$, where B is that set such that $b \in B$ iff $v_{Q'}(B) = b$ for some model $A' = (A, I, s')$ such that s' is identical with s except perhaps at x.

The definitions with regard to validity and completeness then grow out of these modifications in an obvious fashion.

We might next attempt to define a canonical model for EQ out of the Lindenbaum algebra of EQ, $EQ_{/\leftrightarrow}$ (and similarly for RQ). Where V is the set of individual variables of EQ, we might define the *canonical model* $C = (EQ_{/\leftrightarrow}, V, c)$, where c is the *canonical assignment function* defined as

---

[4]The results for EQ were essentially obtained in Anderson and Belnap 1963. They were algebraized and extended to RQ in Belnap 1965. In Belnap 1965 the results were also extended to a slightly larger class of formulas.

follows: $c(x) = x$, and where F is an n-ary predicate, $c(F)$ is that function which takes an n-tuple of individual variables $(x_1, \ldots, x_n)$ into $|Fx_1 \ldots x_n|$. It $^{112}\|_{113}$ is then easy to see that for any formula A, $v_C(A) = |A|$. This is because $EQ_{/\leftrightarrow}$ is an icdl in which $\bigvee_{x \in V} |Ax| = |(Ex)Ax|$.

But now the difficulties arise, because $EQ_{/\leftrightarrow}$ is not a *complete* icdl, and so our "canonical model" is not really a model at all. Although some generalized meets and joins exist in $EQ_{/\leftrightarrow}$ (for example, $\bigvee_{x \in V} |Ax|$ as above), the generalized meets and joins of arbitrary sets of elements need not exist. In fact, a generalized join is known to exist only if it corresponds to an existentially quantified statement, and a generalized meet is known to exist only when it corresponds to a universally quantified statement. The trick would then seem to be to embed $EQ_{/\leftrightarrow}$ (and $RQ_{/\leftrightarrow}$) into a complete icdl, preserving any generalized meets and joins as should happen to exist. Thus we would extend our fake "canonical model" to a genuine canonical model, but unfortunately we do not know how to do such an embedding.

The corresponding difficulty of course arises with the classical predicate calculus, but there it may be overcome, at least for the typical case where the predicate calculus is formulated with at most denumerably many individual variables. The theorem proven is that any Boolean algebra can be embedded in a complete field of sets in such a fashion that the embedding preserves some selected, at most denumerable set of generalized $^{113}\|_{114}$ meets and joins.[5] The proof of this theorem relies heavily upon the deep topological representations of Boolean algebras, and since we do not have such machinery at our disposal for icdl's, it is not clear how, or if, this theorem can be mimicked in any way.

We remark that the various embeddings of icdl's to be found in Chapters II and V (all of them based in essence upon Theorem II:2.1) do not suffice, even though it is in general easy to see that they are embeddings into complete icdl's. The reason is that there is no guarantee that the embeddings preserve generalized meets and joins.

We further remark that even an embedding into a complete icdl which preserves generalized meets and joins would not be sufficient to allow us to go on to give deeper completeness results with respect to M_0 and D. To utilize Theorems II:3.1 and III:3.1 in a way analogous to the implicit use of Theorems II:1.1 and III:1.1 in the proof of Theorem 5.1, we would have to be guaranteed that in the extension of the Lindenbaum algebra to a complete icdl we could find for any two distinct elements, a complete and completely prime filter which separates them. By Theorem II:4.1, this means that the extension must be such that every element is a generalized

---

[5]Rasiowa and Sikorski 1963, pp. 86–88. Cf. also Sikorski 1964, p. 195, and the references given there.

join of completely join-irreducible elements. Thus what is left to prove is made very specific. [114] $\|_{115}$

VIII. Free De Morgan Lattices, Free ICDL's, and the Consistency of First Degree Entailments

It is well-known that the Lindenbaum algebra of the classical propositional calculus formulated with n propositional variables is the free Boolean algebra with n free generators $FB(n)$, where the free generators are the equivalence classes determined by the propositional variables (cf. Chapter I, Section 4). We shall show that the Lindenbaum algebra of the zero degree formulas of E (or R) is both a free De Morgan lattice and a free icdl.

Let us first refer to the system B of the first degree entailments of E. Belnap 1960, pp. 92–93, showed that a formula is provable in this system just when it is a provable first degree entailment in E, and his proof may be easily adapted to show that B also contains just the first degree entailments of R (which shows that E and R agree on their first degree entailment fragments; cf. Belnap 1965). The system B has the following axioms and rules (this is essentially the formulation of Belnap 1960):

B.1 $A \to A$

RB.2 From $A \to B$ and $B \to C$ to infer $A \to C$

B.3 $A \,\&\, B \to A$

B.4 $A \,\&\, B \to B$

RB.5 From $A \to B$ and $A \to C$ to infer $A \to B \,\&\, C$ [115]||[116]

B.6 $A \to A \vee B$

B.7 $B \to A \vee B$

RB.8 From $A \to C$ and $B \to C$ to infer $A \vee B \to C$

B.9 $A \,\&\, (B \vee C) \to (A \,\&\, B) \vee C$

RB.10 From $A \to B$ to infer $\overline{B} \to \overline{A}$

B.11 $A \to \overline{\overline{A}}$

B.12 $\overline{\overline{A}} \to A$

For zero-degree formulas A and B, let us define a relation of co-entailment that holds between them just when both $A \to B$ and $B \to A$ are provable in B. It is trivially verified that such a relation is a genuine equivalence relation that furthermore preserves negation and disjunction in the sense

69

of the Replacement Theorem (which theorem can be proven for B by a trivial application of the argument used by Ackermann 1956 to show his *Ersetzungstheorem* for his system Π'). Thus we may define a Lindenbaum algebra upon the zero degree formulas of B. Letting $|A|$ be the class of formulas that co-entail A, we may define operations upon these congruence classes as follows:

$$|A| \vee |B| = |A \vee B|$$

$$\mathrm{N}|A| = |\overline{A}|$$

We may also introduce a relation by defining

$$|A| \leq |B| \text{ iff } A \to B \text{ is provable in B.}$$

It is now easily verified that this Lindenbaum algebra of B (let us call it $B_{/\leftrightarrow}$) is an icdl (cf. Chapter VII), and hence *a fortiori* a De Morgan lattice. [116]‖[117]

We next observe that it is trivially shown that the system B is consistent with respect to the class of De Morgan lattices, which is to say that if a formula is provable in B then it is De Morgan valid (cf. Section 2, Chapter VII). *A fortiori* the same holds with respect to the class of icdl's. Inspection of the axioms and rules of B shows that they are almost designed for such a proof. For example, B.6, $A \to A \vee B$, is obviously valid since $v_Q(A \vee B) = v_Q(A) \vee v_Q(B)$, and $v_Q(A) \leq v_Q(A) \vee v_Q(B)$. The validity of the other axioms is as obvious, and it is just as obvious that the rules preserve validity. This, together with the previously mentioned fact that B contains the first degree entailments of E (and R), gives us

Theorem 1. *A first degree entailment is provable in E (or R) only if it is valid in the class of De Morgan lattices (and hence in the class of icdl's).*[1]

Using this theorem, we next show

Theorem 2. *The Lindenbaum algebra of the zero degree formulas of E (or R) formulated with n propositional variables is the free De Morgan lattice with n free generators, as well as the free icdl with n free generators.*

Consider any formulation of E (or R) with n propositional variables. Clearly the congruence classes determined by the propositional variables generate the Lindenbaum [117]‖[118] algebra of the zero degree formulas. Consider any mapping f of these generators into an arbitrary De Morgan lattice

---

[1]This was first shown in effect for icdl's in Anderson and Belnap 1963, and our extension to De Morgan lattices is but a trivial generalization.

A. This mapping induces an assignment function s upon the propositional variables so that $s(p_i) = f(|p_i|)$. We thus have a model $Q = (A, s)$ and a valuation determined by the model v_Q. Define $h(|A|) = v_Q(A)$. If h is a well-defined (single-valued) mapping, it is trivially a homomorphism of the Lindenbaum algebra into A. Thus $h(|A| \vee |B|) = h(|A \vee B|) = v_Q(A \vee B) = v_Q(A) \vee v_Q(B) = h(|A|) \vee h(|B|)$, and similarly for N. That h is a well-defined mapping follows from Theorem 1, for suppose that $|A| = |B|$. Then $A \to B$ and $B \to A$ are both provable in E (or R), which means that $v_Q(A) = v_Q(B)$ so that $h(|A|) = h(|B|)$. We remark that from this proof that the Lindenbaum algebra is a free De Morgan lattice, it trivially follows that it is also a free icdl. Thus since we have shown that any mapping of its generators into an arbitrary De Morgan lattice can be extended to a homomorphism, this holds true in particular when the De Morgan lattice is an icdl, and we have already observed that the Lindenbaum algebra is itself an icdl, which completes the theorem.

Theorem 2 not only identifies (up to isomorphism) the free De Morgan lattice with n free generators and the free icdl with n free generators, but simultaneously gives an affirmative answer to the question of their existence. From this identification of free De Morgan lattices and $^{118}\|_{119}$ free icdl's, and from the easily seen fact that every De Morgan lattice with n elements is a homomorphic image of the free De Morgan lattice with n free generators, we obtain immediately the following connection between De Morgan lattices and icdl's:

Theorem 3. *Every De Morgan lattice is a homomorphic image of an icdl.*

We next show

Theorem 4. *The free De Morgan lattice with n free generators $FDML(n)$ is lattice isomorphic to the free distributive lattice with $2n$ free generators $FDL(2n)$.*

The proof of this theorem comes *via* the fact that a necessary and sufficient condition for a set of generators G to be free in a distributive lattice is that for all subsets X and Y of G, $\bigwedge X \leq \bigvee Y$ implies $X \cap Y \neq \emptyset$. Belnap 1965 has shown similarly that a necessary and sufficient condition for a set of generators G to be free in an icdl is that for all subsets X and Y of $G \cup NG$, $\bigwedge X \leq \bigvee Y$ implies $X \cap Y \neq \emptyset$. *Via* Theorem 2 this holds for De Morgan lattices as well. Since it is easy to show by the usual normal form methods that every element in a De Morgan lattice can be represented as a meet of joins of generators or their De Morgan complements,[2] it is

---

[2]Cf. Anderson and Belnap 1961 for a normal form treatment of the zero degree formulas of E.

easy to see that if G freely generates FDML(n) as a De Morgan lattice, then $G \cup NG$ freely generates $^{119}\|_{120}$ FDML(n) as a distributive lattice. But then, recalling our initial remarks, FDML(n) is the free distributive lattice on $2n$ free generators.

Since Birkhoff 1948, p. 146, has shown that the free distributive lattice with n free generators has no more than 2^{2^n} elements, we get immediately from Theorem 4:

Theorem 5. *The free De Morgan lattice with n free generators FDML(n) has no more than $2^{2^{2n}}$ elements.*

This means that no De Morgan lattice with n generators has more than $2^{2^{2n}}$ elements since it may easily seen to be a homomorphic image of FDML(n).

We remark that in virtue of the fact that the word problem for FDL(n) is solvable,[3] that Theorem 5 implies that the first degree entailment fragments of E and R are decidable (a fact first shown in Anderson and Belnap 1961). $^{120}\|_{121}$

---

[3]Cf. Whitman 1961 for terminology and Birkhoff 1948, pp. 145–146 for proof.

IX. An Intuitive Semantics for First Degree Entailments

This chapter is intended as a tentative step toward an intuitive (as opposed to algebraic) semantics for relevant implication. The motivating idea is that A should relevantly imply B iff, in some appropriate sense, whatever B is about, A is about as well. This is meant as an explication of the classical metaphor of the containment of the predicate in the subject of an analytic truth (cf. Kant's *Critique of Pure Reason*, B 10). We now set ourselves the task of eliciting the appropriate sense of 'about'.

We first observe that if we want A to entail[1] $A \vee B$ (as we do), then we must be using the word 'about' in a rather special sense. For to use an example from Nelson Goodman's careful 1961 paper "About," 'Florida is Democratic' entails 'Florida or Maine is Democratic,' and yet the second statement seems to be about Maine, whereas the first does not. We are forced to maintain that *really* the second $^{121}\|_{122}$ statement is not about Maine in the sense we are after.[2] Our sense of 'about' thus differs significantly from Goodman's, since he admits that the second statement is about Maine (which is the natural thing to do) and then proceeds to patch up his initially plausible claim that if A entails B and B is about something, then A is about that thing as well.[3]

We can justify our conclusion that in some sense the second statement is not *really* about Maine by observing that the second statement does not really give us any definite information about Maine (nor does it give us the definite sort of information about Florida that the first statement does).

---

[1] In the sequel we shall talk about entailment, rather than relevant implication, mainly because "A relevantly implies B" is such an awful, artificial thing to say. This slippage is harmless, since all of our results will pertain, to first degree relevant implications only, and the system E of entailment and the system R of relevant implication agree on their first degree implication theorems.

[2] Alternatively, we could conclude that the first statement is *really* about Maine after all. But as Goodman points out, such a move would lead to the absurd conclusion that any statement is about everything, since we can always add disjuncts appropriately.

[3] Goodman has a goal complementary to ours; he wants to explicate *aboutness* is terms of entailment. In this connection, it is worth pointing out that some of the difficulties that arise for Goodman are due to his meaning by 'entailment,' (provable) material implication. If he were to mean by 'entailment,' entailment (in the system E), many of these difficulties would be obviated. Thus, for example, Goodman is forced to maintain that a tautology is not about anything (roughly, because it is implied by any statement), but, of course, a tautology is not *entailed* by any statement.

The idea is that a disjunction gives us definite information about something only if each disjunct gives us definite information about that thing as well. This assumption can be motivated on $^{122}\|_{123}$ information-theoretical grounds, since it is universally assumed that a disjunction can contain no more information than either disjunct.[4] So if some disjunct gives us no definite information about something, we cannot expect the disjunction as a whole to give us definite information about that thing. We assume conversely that if each disjunct gives us definite information about something, then the disjunction as a whole gives us definite information about that same thing. Our assumptions concerning conjunction are (dually) in accord with those concerning disjunction, and perhaps sound more obvious. We assume that a conjunction gives us definite information about something iff either conjunct gives us definite information about that thing (consider that a conjunction must contain at least as much information as either conjunct).

We have treated disjunction and conjunction, but what about negation? If A gives us definite information about something, does $\overline{A}$ give us definite information about the same thing? The answer is that it depends. Suppose that p is an "atomic" statement and that p gives us definite information about a certain topic x (for purposes of illustration, imagine that p is 'Florida is Democratic' and x is Florida). Then surely $\overline{p}$ gives definite information about x as well (as 'Florida is *not* $^{123}\|_{124}$ Democratic' gives definite information about Florida). So if A is atomic and A gives us definite information about x, so does $\overline{A}$. But the story gets complicated when A is not atomic. Suppose that q is an atomic statement independent from p relative to the topic x (thus q does not give definite information about x; q might be 'Snow is white,' which presumably gives no definite information about Florida). If A is $p \& q$, then $\overline{A}$ is $\overline{p} \vee \overline{q}$, which gives us no definite information about x, although A does. Or conversely, if A is $\overline{p} \vee \overline{q}$, then $\overline{A}$ is $p \& q$, which gives us definite information about x, although A does not. Or if A is $(p \vee q) \& q$, then $\overline{A}$ is $(\overline{p} \& \overline{q}) \vee \overline{q}$, and neither A nor $\overline{A}$ gives us definite information about x.

In general then, given what A gives definite information about, we cannot tell what $\overline{A}$ gives definite information about, and conversely. Since understanding the meaning of a sentence involves understanding the meaning of its negation as well, any adequate explication of the meaning of a sentence in terms of what it gives definite information about must bring out as well what its negation gives definite information about.

This informal discussion was intended to motivate the notion of an interpretation of a formula in terms of an "aboutness" valuation, which assigns in some inductive fashion to each formula a pair of classes, the first class to be

---

[4]Cf. Carnap and Bar-Hillel 1952, p. 16.

thought of as the things that the formula gives definite information about, and the second class to be [124]||[125] thought of as the things that its negation gives definite information about. The reader may have noticed that we have been vague as to what sort of thing a sentence is about. Is the sentence 'Florida is Democratic' about Florida, the Democratic Party, the class consisting of both, the pair consisting of both, the proposition that Florida is Democratic, or even, as Nuel D. Belnap, Jr. has suggested, the question 'Is Florida Democratic?', or what? Fortunately, these problems can be avoided by simply saying that the sentence is about a certain topic or *topics*. This answer will not be satisfactory for those who want to know what sentences are about *ultimately*. However, this answer, together with the inductive rules governing "aboutness" valuations, suffices for determining the validity of first degree entailments. The situation is perhaps comparable to the situation with regard to the role of truth in the semantics of the classical propositional calculus. One can imagine logicians with radically conflicting metaphysical theories of truth agreeing that whatever truth is *ultimately*, it must behave formally the way Tarski's notion of a valuation says it does. Of course, one can also imagine logicians whose metaphysics of truth will not allow them to accept the Tarskian notion of a valuation (cf. Łukasiewicz and 3-valued logic), but a formal explication cannot please everyone. We hence proceed without further argument to give a formal explication of an "aboutness" valuation. [125]||[126]

Let us say that Q is an *aboutness model* iff Q is a pair (X, s), where X is a set (called the *universe of discourse*) and s is a function assigning to each propositional variable p a pair (X_1, X_2) of subsets of X, to be thought of, respectively, as the topics that p gives definite information about and the topics that $\overline{p}$ gives definite information about. Let us call each pair of subsets of X a *proposition surrogate*, as seems appropriate since such a pair when assigned to a formula gives a partial representation of the meaning of the formula. It is easy to define operations upon the set of all proposition surrogates of a given universe of discourse X so as to make it a De Morgan lattice that has, as we shall see, some intuitive appeal. Thus define $N(X_1, X_2) = (X_2, X_1)$; $(X_1, X_2) \vee (Y_1, Y_2) = (X_1 \cap Y_1, X_2 \cup Y_2)$; and $(X_1, X_2) \wedge (Y_1, Y_2) = (X_1 \cup Y_1, X_2 \cap Y_2)$. Note then that $(X_1, X_2) \leq (Y_1, Y_2)$ iff $Y_1 \subseteq X_1$ and $X_2 \subseteq Y_2$. Let us call this *the De Morgan lattice of proposition surrogates of the universe of discourse X*.

Given an aboutness model $Q = (X, s)$, we can define a (zero degree) *aboutness valuation* determined by Q, v_Q, as the (zero degree) valuation determined by the De Morgan lattice of all proposition surrogates of X with the assignment function s (remembering that we have already defined such a valuation for De Morgan lattices in general in Chapter VII). It is easily seen that this definition is equivalent to the following conditions: for all zero [126] degree formulas A,

if A is a propositional variable, $v_Q(A) = s(A)$;

if A has the form $\overline{B}$, then if $v_Q(B) = (X_1, X_2)$, $v_Q(A) = (X_2, X_1)$;

if A has the form $B \vee C$, then if $v_Q(B) = (X_1, X_2)$ and $v_Q(C) = (Y_1, Y_2)$, $v_Q(A) = (X_1 \cap Y_1, X_2 \cup Y_2)$.

if A has the form $B \,\&\, C$, then if $v_Q(B) = (X_1, X_2)$ and $v_Q(B) = (Y_1, Y_2)$, $v_Q(C) = (X_1 \cup Y_1, X_2 \cap Y_2)$.

It is easy to see that these conditions are in accord with our informal motivating remarks.[5] The only thing that has been added are rules for computing what the negation of a disjunction (or conjunction) gives definite information about, given what the negation of each disjunct (or conjunct) gives definite information about. Note that these additional rules are intuitively sound. For example, it is right that in the condition determining the value of a disjunction we intersect X_2 and Y_2, because $\overline{A}$ is $\overline{B} \,\&\, \overline{C}$, which gives definite information about every topic that $\overline{B}$ gives definite information (the members of X_2) as well as every topic that $\overline{C}$ gives definite information about (the members of Y_2).

We can now define an intuitive notion of validity by a simple application of our general definition of [127] validity with respect to a class of De Morgan lattices in Chapter VII, specifying the class in question to be the class of De Morgan lattices of proposition surrogates of universes of discourse. We accordingly say that a first degree entailment $A \to B$ is *true in an aboutness model* $Q = (X, s)$ iff when $v_Q(A) = (X_1, X_2)$ and $v_Q(B) = (Y_1, Y_2)$, then $Y_1 \subseteq X_1$ and $X_2 \subseteq Y_2$, and otherwise as *false in the model*. This definition says that a statement 'A entails B' is true iff every topic that B gives definite information about, A gives definite information about as well, and every topic that $\overline{A}$ gives definite information about, $\overline{B}$ gives definite information about as well.[6]

One can easily imagine that an entailment might be true in a given aboutness model, although not true in all aboutness models. Thus, for example, the statement 'John is a bachelor entails John is unmarried' would surely be true in an aboutness model properly associated with the English language, but we can imagine that the same statement would be false in most

---

[5] Note also how these values for disjunctions and conjunctions are analogous in their behavior on the first components to *information content*, in Carnap and Bar-Hillel 1952, p. 16. For, $\mathrm{Cont}(A \vee B) = \mathrm{Cont}(A) \cap \mathrm{Cont}(B)$, and $\mathrm{Cont}(A \,\&\, B) = \mathrm{Cont}(A) \cup \mathrm{Cont}(B)$.

[6] The last clause, which relates what $\overline{A}$ and $\overline{B}$ gives definite information about, may be motivated by reflecting on the fact that if A entails B, then $\overline{B}$ entails $\overline{A}$.

other aboutness models, which would, so to speak, assign the same symbols (except for the word 'entails') different "meanings."[7] [128]$\|_{129}$

So we next define a first degree entailment as *valid in a universe of discourse* X iff it is true in every aboutness model $Q = (X, s)$, and otherwise as *falsifiable in* X.

We finally define a first degree entailment as *valid* (simpliciter) iff it is valid in every universe of discourse, and otherwise as *falsifiable*.

It is interesting that the De Morgan lattice D, which played a central role in the algebraic semantics of first degree entailments of Chapter VII, can be represented as the De Morgan lattice of proposition surrogates of a universe of discourse consisting of but a single topic x, as is shown in the following diagram.

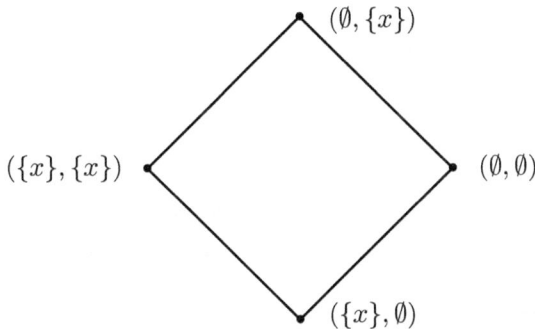

Because of this, *via* the fact that we have already proven that a first degree entailment in provable in E (or R) iff it is valid in the De Morgan lattice D (combining Theorems VII:5.1 and VIII:1), we get immediately [129]$\|_{130}$

Theorem 1. *A first degree entailment is provable in E (or R) iff it is valid in a universe of discourse consisting of just a single topic.*

And this theorem, together with the fact that we have already proven that a first degree entailment is provable in E (or R) only if it is valid with respect to the class of De Morgan lattices in general (Theorem VIII:1), gives us

Theorem 2. *A first degree entailment is provable in E (or R) iff it is valid.*

---

[7]These remarks are not intended to convey the impression that we believe that there is a hard and fast distinction between "descriptive" phrases, such as 'John is a bachelor,' and "logical constants," such as 'entails.' But such a distinction can be made *relative* to a given semantics.

It is interesting to note that every De Morgan lattice is isomorphically embeddable in a De Morgan lattice of proposition surrogates of some universe of discourse X. The proof is by observing that any De Morgan lattice of proposition surrogates of a universe of discourse X is isomorphic to a 2-product of the field of all subsets of X, under the following mapping h: $h(X_1, X_2) = (\overline{X_1}, X_2)$. This observation leads via Theorem VI:3 immediately to the desired result.

It is interesting to note that the fact that the mapping h defined above is an isomorphism reflects the possibility that we could have just as well assigned to each formula A a pair (X_1, X_2), where X_1 is the set of topics that A is about and X_2 is the set of topics that $\overline{A}$ is about, and we would have assigned the formula just as much information as to what it is about as we do in our official way. We avoided this because this mixing of $^{130}\|_{131}$ unaboutness with aboutness gets conceptually confusing. But it would have been algebraically neater, since then the appropriate intuitive definition of an aboutness valuation v_Q would have been a natural mapping into a 2-product of the field of all subsets of the topic set, taking v into joins and & into meets. Thus, for example, when $v_Q(B) = (X_1, X_2)$ and $v_Q(C) = (Y_1, Y_2)$, then $v_Q(B \vee C) = (X_1 \cup Y_1, X_2 \cup Y_2)$, which is more symmetrical looking than our official way of defining valuations.

In this connection it is perhaps worth pointing out that a proposition surrogate (X_1, X_2) of a topic set X can be represented by a pair $(\{a_{1_x}\}_{x \in X}, \{a_{2_x}\}_{x \in X})$ where each $a_{i_x} \in \mathbf{2}$. Thus by a coding of "in" and "out" with 1 and 0 we can represent the surrogate. We can let $a_{1_x} = 1$ mean that $x \in X_1$ and $a_{2_x} = 1$ mean that $x \in X_2$. Then if we define N as a rotation, and meets and joins in the "mixed" fashion, all of this done in the obvious parallel to the way it was done with proposition surrogates, we obtain a De Morgan lattice that is related to a 2-product of a direct product of $\mathbf{2}$ in the same way that a De Morgan lattice of proposition surrogates is related to a 2-product of a field of sets.

Now that we have seen what De Morgan lattices and 2-products are all "about," one naturally wonders what icdl's and 3-products are all about. Remembering that the last component of an element of a 3-product behaved $^{131}\|_{132}$ in a classical "on-off" manner (for example, the last component of an element of a 3-product of a direct product of $\mathbf{2}$ was either all 0's or all 1's), whereas the first two components behaved like the components of an element of a 2-product, we remark that a 3-product should probably be thought of as a proposition surrogate together with a truth value. Such entities are necessary for defining validity for formulas that do not have the form of an entailment (cf. Belnap 1965 for a definition of validity for first degree formulas (truth functions of truth functions and first degree entailments) in terms of icdl's and corresponding completeness results).

We may end this chapter by suggestively claiming that the problem of an intuitive semantics for the whole of relevant implication (R) consists of deciding, from a knowledge of what A is about and what B is about, what $A \rightarrow B$ is about. [132]$\|$[133]

X. THE ALGEBRAIC STRUCTURE OF R

1. The Operation of Relevant Implication

We have seen in Chapter VII that the Lindenbaum algebra of R is an icdl, and we have seen in Chapter VIII that, indeed, the Lindenbaum algebra of the zero degree formulas of R is a free icdl. But these observations contribute little to elucidating the algebraic structure of the whole of R. The difficulty is that there is no operation in icdl's that corresponds to relevant implication. Relevant implication can be represented in icdl's only as a relation. So that an axiom of R like A.3, $A \rightarrow B \rightarrow .B \rightarrow C \rightarrow .A \rightarrow C$, can be only imperfectly represented in icdl's. Icdl's do represent the transitivity of the implication relation that is a consequence of this axiom, since the rule of transitivity—from $A \rightarrow B$ and $B \rightarrow C$ to infer $A \rightarrow C$—is represented by the transitivity of the partial-ordering relation. But for an algebraic structure to represent this axiom directly it would have to have an "arrow" operation such that for elements a, b, c, $a \rightarrow b \leq (b \rightarrow c) \rightarrow (a \rightarrow c)$, just as in the Lindenbaum algebra of R, $|A| \rightarrow |B| \leq (|B| \rightarrow |C|) \rightarrow (|A| \rightarrow |C|)$.

It turns out that a certain kind of residuated lattice-ordered semi-group has such an "arrow" operation $^{133}\|_{134}$ to represent relevant implication.[1] The reader is warned that no "deep" theorems concerning either R or the related special kind of semi-groups will be forthcoming in the sequel. There is, at least in the present stage of investigation, no "spin off" of the sort associated with the McKinsey–Tarski identification of the study of S4 with the study of closure algebras. Thus the identification of the study of R with the study of the special kind of semi-group does not magically solve any of the unsolved problems concerning R; nor does this identification yield any significant theorems concerning the special kind of semi-groups. In particular, the decidability of both theories is left open.

The subsequent theorems then should be of no interest to the logician *qua* logician, nor to the mathematician *qua* mathematician; but hopefully they may be of interest to the logician *qua* mathematician or his dual (both of which fall under the common abstraction of the mathematical logician). Needless to say, we hope ultimately that the connection between R and the special kind of semi-group will prove of real use in the study of both since, in mathematics, two points of view are often better than one. $^{134}\|_{135}$

---

[1]The author wishes to thank Robert K. Meyer for having suggested to him a connection between relevant implication and residuation. Meyer discusses certain analogies in his doctoral dissertation of 1966.

2. Preliminaries on Lattice-Ordered Semi-Groups

Let us recall that a *semi-group* $(S, \circ)$ is a non-empty set S that is closed under a binary operation $\circ$ that is associative, i.e., $a \circ (b \circ c) = (a \circ b) \circ c$. We say of a semi-group that it is commutative if $ab = ba$.[2] A semi-group is *with identity* if it has an element e (*the identity*) such $ea = ae = a$ (recall that e is provably unique). A semi-group with identity is a *group* if for every element a there exists an element a^{-1} (*the inverse* of a) such that $aa^{-1} = a^{-1}a = e$ (recall that inverses are provably unique).

We say of a semi-group that it is *lattice-ordered* (briefly, that it is an *l-semi-group*) if it is a lattice as well (though neither the meet nor the join need be the same as the multiplication operation $\circ$), and $a(b \vee c) = (ab) \vee (ac)$ and $(b \vee c)a = (ba) \vee (ca)$. Note that it follows from this that $a \leq b$ implies $ca \leq cb$ and $ac \leq bc$ (cf. Certaine 1943, p. 39). Note in particular that the l-semi-group may be a group, in which case we call it an *l-group*.[3] $^{135}\|_{136}$

We say of an l-semi-group that it is *right-residuated* if for every pair of elements a, b there exists an element $a : b$ (called the *right residual* of a by b) such that $x \leq a : b$ iff $xb \leq a$. Clearly $a : b$ is the greatest element x such that $xb \leq a$, and indeed is the join of all such elements x. Similarly, *left-residuation* is defined so that for every pair of elements a, b there exists an element $a :: b$ such that $x \leq a :: b$ iff $bx \leq a$. If an l-semi-group is commutative, then it is right-residuated iff it is left-residuated. If an l-semi-group is both right- and left-residuated, then we simply say that it is *residuated*.[4]

Residuation is a common abstraction of the ideal quotient in ring theory and of division in number theory. For examples of the former cf. Fuchs 1963, p. 190. For an instructive example of the latter consider the multiplicative group of the positive rationals $\mathbb{Q}^+$ ordered so $a \leq b$ iff a integrally divides b, i.e., b/a is a positive integer. This is a residuated l-group in which $a : b = a/b$ (cf. Birkhoff 1948, p. 202). Indeed Certaine 1943 has shown that any l-group is residuated and that $a : b = ab^{-1}$ and $a :: b = b^{-1}a$ (p. 61). $^{136}\|_{137}$

---

[2] Note that we permit ourselves to drop the '$\circ$' notation for multiplication in favor of juxtaposition when convenient.

[3] The concept of an l-semi-group should probably be credited to Ward and Dilworth 1939 (although they assume commutativity and the existence of an identity). Birkhoff 1942 drops the commutativity assumption (although he still retains the identity assumption) and calls the resulting structure a "groupoid." Certaine 1943, in his classic study, follows Birkhoff's usage. The reader should be warned that many modern writers mean by a groupoid simply a set with a binary operation (with no assumptions even about associativity, let alone about the existence of an identity), as in Fuchs 1963. The reader should be further warned that Birkhoff 1948 means by a "semi-group" what we have called a semi-group with identity and what he called in 1942 a "groupoid." Our terminology, perhaps more modern, derives from Fuchs 1963.

[4] The concept of residuation, although as old as ideal quotients and division, was probably first abstractly formulated by Ward and Dilworth 1939.

An especially instructive example of a residuated l-semi-group, from a logical point of view, is a Boolean algebra in which meet is taken as the multiplicative operation and $a : b = \bar{b} \vee a$. Similarly instructive is a pseudo-Boolean algebra in which meet is taken as the multiplicative operation and $a : b$ is the pseudo-complement of b relative to a. Both of these examples can be found in Certaine 1943, p. 61.

We now gather some useful properties of commutative l-semi-groups. Since we are, dealing with the commutative case we need not distinguish between left- and right-residuals and shall denote both by $a : b$.

P1. $a \leq b$ implies $a : c \leq b : c$.

P2. $b : a \leq (b : c) : (a : c)$.

P3. $a \leq b$ implies $c : b \leq c : a$.

P4. $b : a \leq (c : a) : (c : b)$.

P5. $ab \leq c$ iff $a \leq c : b$ iff $b \leq c : a$.

P6. $c : (ab) = (c : b) : a = (c : a) : b$.

P7. $a(b : a) \leq b$.

P8. $a \leq b : (b : a)$.

P9. $a \leq (ab) : b$.

P10. If $\bigwedge_{x \in X} a_x$ exists, then so does $\bigwedge_{x \in X} (a_x : b)$ and $\bigwedge_{x \in X} (a_x : b) = (\bigwedge_{x \in X} a_x) : b$.

P11. If $\bigvee_{x \in X} a_x$ exists, then so does $\bigwedge_{x \in X} (b : a_x)$ and $\bigwedge_{x \in X} (b : a_x) = b : (\bigvee_{x \in X} a_x)$.

P12. $(a \wedge b)(a \vee b) \leq ab.$ [137]$\|$[138]

Proofs of properties P1–11 may be found in Certain 1943, pp. 68–69. P12, in the presence of commutativity, follows immediately from Certaine's 3) on p. 39. We also have

P13. $a \leq c$ and $b \leq d$ imply $ab \leq cd$.

Proof: $a \leq c$ implies $ab \leq cb$, but $b \leq d$ implies $cb \leq cd$, and hence by transitivity $ab \leq cd$.

If we assume that the l-semi-group has an identity e, we get the following further properties (proven in Certaine 1943, pp. 69–70).

P14. $a : e = a$.

P15. $e \leq a : a$.

P16. $a \leq b$ implies $e \leq b : a$.

If we further assume that the l-semi-group is *upper semi-idempotent* or *square increasing*, i.e., that $a \leq aa$, we get the following.

P17. $a \wedge b \leq ab$.

Proof: 1) $(a \wedge b)(a \vee b) \leq ab$. P12.
2) $a \wedge b \leq a \wedge b$ and $a \wedge b \leq a \vee b$. Lattice properties.
3) $(a \wedge b)(a \wedge b) \leq ab$. From 1) and 2) *via* P13.
4) $a \wedge b \leq ab$. From 3) by upper semi-idempotency.

P18. $(b : a) : a \leq b : a$.

Proof: 1) $a((b : a) : a) \leq b : a$. P7.
2) $aa((b : a) : a) \leq b$. From 1) by P5.
3) $a((b : a) : a) \leq b$. From 2) by P13 since $a \leq aa$.
4) $(b : a) : a \leq b : a$. From 3) by P5. [138][139]

It might be anticipated that we are about to define a *De Morgan lattice-ordered semi-group* $(S, \circ, \leq, N)$. We shall, of course, want $(S, \circ, \leq)$ to be a lattice-ordered semi-group and $(S, \leq, N)$ to be a De Morgan lattice. But we should further like N to have some direct interaction with $\circ$. Let us then require the following:

A) $a \circ b \leq c$ iff $b \circ Nc \leq Na$, and $a \circ b \leq c$ iff $Nc \circ a \leq Nb$.

Property A) will ultimately be motivated by the applications to R that are forthcoming in the next section. But we can even at this stage attach some algebraic meaning to it by observing that it holds both in Boolean algebras treated as l-semi-groups (where N is Boolean complementation) and in l-groups (where N is taken as inversion). Indeed, under these interpretations of N both Boolean algebras and l-groups turn out to be De Morgan lattice-ordered semi-groups. It was observed in Birkhoff 1942 that inversion is a dual automorphism of period two on the underlying lattice of an l-group, and that the underlying lattice is necessarily distributive (cf. pp. 301 and 306). Kalman 1958 offered De Morgan lattices that are "normal" in his sense (cf. Chapter IV) as a "common abstraction of Boolean algebras and l-groups," thus seemingly presenting them as an answer to problem 105 of Birkhoff 1948: "Is there a common abstraction which includes Boolean algebras (rings) and l-groups as special cases?" Unfortunately, De Morgan lattices by themselves take no account of the multiplicative operation of the group. We $^{139}\|_{140}$ therefore suggest that our De Morgan lattice-ordered semi-groups, in this respect at least, provide a better answer to Birkhoff's problem; this answer could be further improved by adding the postulate that the De Morgan lattice is "normal" in the sense of Kalman, as well as the postulate that the semi-group has an identity.

It is interesting to note the strength of postulate A). Thus it gives us

Theorem 1. *Every De Morgan lattice-ordered semi-group is residuated, with $a : b = N(b \circ Na)$ and $a :: b = N(Na \circ b)$.*

We show the first identity by using the first part of A). Thus $x \circ b \leq a$ iff $b \circ Na \leq Nx$. But since N is an involution, this last holds iff $x \leq N(b \circ Na)$. So $x \circ b \leq a$ iff $x \leq N(b \circ Na)$, and the first identity follows immediately from the definition of $a : b$. The second identity follows similarly from the second part of A).

It is interesting to note that an operation $+$ dual to $\circ$ can be defined via $a + b = N(Na \circ Nb)$. This operation is associative, but distributes over meet (instead of join) and satisfies the dual of A), namely, $a + b \geq c$ iff $b + Nc \geq Na$, and $a + b \geq c$ iff $Nc + a \geq Nb$. We may then define operations of right-difference $(\dot{-})$ and left-difference $(\ddot{-})$. They may be defined by $x \geq a \dot{-} b$ iff $a \leq b + x$, and $x \geq a \ddot{-} b$ iff $a \leq x + b$, and it may be shown that $a \dot{-} b = N(b : a) = a \circ Nb = N(Na + b)$, and similarly for $a \ddot{-} b$. All of this suggests an obvious equivalent (but dual) formulation of De Morgan lattice-ordered semi-groups, in which $+$ is taken as primitive and $\circ$ defined in terms of it. $^{140}\|_{141}$

Let us now define a special kind of De Morgan lattice-ordered semi-group that will be useful in the sequel. Since we shall no longer be talking of De Morgan lattice-ordered semi-groups in general, we shall not risk confusion if we call this special kind a *De Morgan semi-group* for the sake of brevity. A *De Morgan semi-group* then is a quadruple $(S, \circ, \leq, N)$ that is a commutative, upper semi-idempotent, De Morgan lattice-ordered semi-group. When the underlying De Morgan lattice is an icdl as well, we shall speak of an *intensional semi-group*.[5]

Evidently, P1–13 and P17–18 hold of any De Morgan semi-group, and P14–16 hold as well if it has an identity. *Via* Theorem 1, we may prove the following two further properties of De Morgan semi-groups.

---

[5]Note that a De Morgan semi-group is *not* a generalization of an l-group, not even a commutative l-group, because of upper semi-idempotence. And an intensional semi-group is even less a generalization, since in any group $e = e^{-1}$. But De Morgan semi-groups and intensional semi-groups are still generalizations of Boolean algebras.

P19. $a : b = Nb : Na$.

Proof: By Theorem 1, $a : b = N(b \circ Na)$ and $Nb : Na = N(Na \circ NNb) = N(Na \circ b)$. But then $N(b \circ Na) = N(Na \circ b)$ by commutativity.

P20. $Na : a \leq Na$.

Proof: By Theorem 1, $Na : a = N(a \circ NNa) = N(a \circ a)$. But $N(a \circ a) \leq Na$ follows from upper semi-idempotency, $a \leq a \circ a$, by the antitonicity of N. [141]$\|_{142}$

A few more definitions, and then we shall be prepared to discuss the algebraic structure of R. Let us define a *homomorphism* between two De Morgan semi-groups, $(S, \circ, \leq, N)$ and $(S', \circ', \leq', N')$, as a De Morgan homomorphism h between $(S, \leq, N)$ and $(S', \leq', N')$ that is also a semi-group homomorphism between $(S, \circ)$ and $(S', \circ')$, i.e., $h(a \circ b) = h(a) \circ' h(b)$. Observe that in virtue of the interdefinability of multiplication and residuation, the condition that h is a semi-group homomorphism is equivalent to the condition that h preserves residuation, i.e., $h(a : b) = h(a) :' h(b)$. A one-to-one homomorphism is an *isomorphism*.

If the two De Morgan semi-groups S and S' have identities e and e', respectively, then if h is a homomorphism between them such that $h(e) = e'$, we shall call h an *identity-preserving homomorphism* (or, *e-homomorphism*). If h is one-to-one, it is an *e-isomorphism*.

In accord with these definitions, we shall define *the free De Morgan semi-group with n free generators a_i* FDMSG(n) as a De Morgan semi-group such that any mapping of the a_i into an arbitrary De Morgan semi-group S can be extended to a homomorphism of FDMSG(n) into S. We shall similarly define *the free De Morgan semi-group with identity with n free generators a_i* FDMSGw/I(n) as a De Morgan semi-group with identity such that any mapping of the a_i into an arbitrary De Morgan semi-group with identity can be extended to an *e*-homomorphism. We may [142]$\|_{143}$ define *free intensional semi-groups* and *free intensional semi-groups with identity* similarly.

3. R and De Morgan Semi-Groups

We may introduce a new connective into R by defining $A \circ B = \overline{A \to \overline{B}}$.[6] We may then define a multiplicative operation upon the Lindenbaum algebra of R, $R_{/\leftrightarrow}$, by letting $|A| \circ |B| = |A \circ B|$ (note the harmless ambiguity of '$\circ$' used both syntactically and algebraically). We now show that $R_{/\leftrightarrow}$ is a De Morgan (indeed, an intensional) semi-group under $\circ$.

We first recall that it was established in Chapter VI that $R_{/\leftrightarrow}$ is an icdl, so it only remains to show that $\circ$ is commutative, associative, upper semi-idempotent, distributive over join, and that it satisfies A). We sketch proofs of these, making free use of replacement and elementary properties of R such as double negation. [143]$\|_{144}$

To show commutativity, it suffices to prove in R $A \circ B \to B \circ A$, i.e., $\overline{A \to \overline{B}} \to \overline{B \to \overline{A}}$, which follows immediately from contraposition.

To show associativity (in the presence of commutativity), it suffices to show $A \circ (B \circ C) \to (A \circ B) \circ C$, i.e., $\overline{A \to \overline{\overline{B \to \overline{C}}}} \to \overline{\overline{A \to \overline{B}} \to \overline{C}}$, which by contraposition is equivalent to $\overline{A \to \overline{B}} \to \overline{C} \to .A \to (B \to \overline{C})$, which in turn is equivalent by another contraposition to $C \to (A \to \overline{B}) \to .A \to (B \to \overline{C})$, which is equivalent by yet another contraposition to $C \to (A \to \overline{B}) \to .A \to (C \to \overline{B})$, which is just an instance of permutation.

To show semi-idempotency, it suffices to prove $A \to (A \circ A)$, i.e., $A \to \overline{(A \to \overline{A})}$, which by contraposition is just $(A \to \overline{A}) \to \overline{A}$, the *reductio* axiom.

To show that $\circ$ distributes over join, it suffices to prove $A \circ (B \lor C) \leftrightarrow (A \circ B) \lor (A \circ C)$, i.e., $\overline{A \to \overline{B \lor C}} \leftrightarrow (\overline{A \to \overline{B}} \lor \overline{A \to \overline{C}})$, which by contraposition is equivalent to $(A \to \overline{B} \& \overline{C}) \leftrightarrow ((A \to \overline{B}) \& (A \to \overline{C}))$, which follows immediately from the conjunction introduction and elimination axioms.

Finally, we show that $\circ$ satisfies property A). Since we have already shown commutativity, obviously it suffices to show only one part of A). And since we have double negation, it then suffices to show $(A \circ B) \to C \to .(B \circ \overline{C}) \to \overline{A}$, i.e., $\overline{A \to \overline{B}} \to C \to .\overline{B \to \overline{\overline{C}}} \to \overline{A}$, which by contraposition is $\overline{C} \to (A \to \overline{B}) \to .A \to (B \to C)$, which by another contraposition is $\overline{C} \to (A \to \overline{B}) \to .A \to (\overline{C} \to \overline{B})$, which is just an instance of permutation. [144]$\|_{145}$

---

[6]This connective is dual to Belnap's 1959 "intensional disjunction," defined as $A \uplus B = \overline{A} \to B$ (though defined for E). It might be called "intensional conjunction" then, but it has so few of the properties of ordinary truth-functional conjunction that to think of it in this way is misleading. Peter Woodruff has pointed out that it is intuitive to read "$A \circ B$" as A is consistent with B, since it asserts that A does not relevantly imply $\overline{B}$. It is thus strictly analogous to Lewis and Langford's 1959 consistency operation defined in terms of strict implication. But unlike their consistency operation, one cannot prove in R the negation of $A \circ (B \& \overline{B})$, i.e., one cannot prove that A is inconsistent with $B \& \overline{B}$, which is something to keep in mind when reading '$\circ$' as a consistency operation.

This completes the proof. Let us observe by 2.1, that the residual of $|A|$ by $|B|$, $|A| : |B|$, is just $|\overline{B \circ \overline{A}}|$, which by definition of $\circ$ (and two double negations) is just $|B \to A| = |B| \to |A|$. Let us record these facts in

Theorem 1. *The Lindenbaum algebra of R, $R_{/\leftrightarrow}$, is a De Morgan (indeed, an intensional) semi-group under $\circ$ (where $|A| \circ |B| = |A \circ B|$, and where $A \circ B$ is defined in turn as $\overline{A \to \overline{B}}$. The residual of $|A|$ by $|B|$, $|A| : |B|$, is $|B| \to |A|$.*

However, the Lindenbaum algebra of R is not a free De Morgan semi-group, not even a free intensional semi-group. This is because $A \to A \to B \to B$ is a theorem of R, but it is not true in intensional semi-groups in general that $b : (a : a) \le b$. The following proof sketch of $A \to A \to B \to B$ is instructive. We start with $A \to A \to B \to .A \to A \to B$ and permute, giving us $A \to A \to .A \to A \to B \to B$. But this last, together with $A \to A$, gives us by *modus ponens* the desired theorem. Intensional semi-groups in general do not take account of such a use of *modus ponens* (which we might express algebraically as: $a \le b$ and $b : a \le d : c$ imply $c \le d$).

For an argument that $b : (a : a) \le b$ is indeed independent of the postulates for an intensional semi-group, consider the intensional semi-group consisting of the elements -1, -0, $+0$, and $+1$, where in the lattice structure $-1 \le -0 \le +0 \le +1$ so it is the four element chain, where $N(\pm a) = \mp a$, and where $\circ$ is defined by the following matrix:[7] [145]$\|_{146}$

$\circ$	-1	-0	$+0$	$+1$
-1	-1	-1	-1	-1
-0	-1	$+1$	$+1$	$+1$
$+0$	-1	$+1$	$+1$	$+1$
$+1$	-1	$+1$	$+1$	$+1$

It is easily verified that this is indeed an intensional semi-group. Then, remembering $a : b = N(b \circ Na)$, $b : (a : a) = N(N(a \circ Na) \circ Nb)$, and upon letting $a = b = +0$, we can compute $b : (a : a) = +1$, but $+1 \nleq +0$. Observe further that $+0 : +0 = -1$, and hence $+0 : +0 \le d : c$ for all c and d, even though it is not the case for all c and d that $c \le d$, so we can see directly how this intensional semi-group violates the algebraic analogue of *modus ponens* mentioned above. It is interesting to observe that if a De Morgan semi-group has an identity e (as this one does *not*), then it must have the

---

[7]This matrix derives from an enlargement of an implication matrix found in Anderson and Belnap's *Entailment* 1965 8.4.1 (matrix IV).

algebraic analogue of *modus ponens*, for by P16, $a \leq b$ implies $e \leq b : a$, but then by transitivity, $e \leq d : c$, and hence by P16, $c \leq d$.

This raises the question of whether $R_{/\leftrightarrow}$ has an identity. Perhaps it is free in intensional semi-groups with identity. Is there then a formula A such that $|A|$ is the identity of $R_{/\leftrightarrow}$? We know by P15 and Theorem 1, that if there is such a formula A, then $|A| \leq |B| \rightarrow |B|$ for every formula B, i.e., $A \rightarrow .B \rightarrow B$ is a theorem of R for every formula B, and in particular, for every propositional variable p, $A \rightarrow .p \rightarrow p$ is a theorem of R. But it may be shown for R $^{146}\|_{147}$ that no formula of the form $C \rightarrow D$ is a theorem unless C and D share a propositional variable. (This is shown by essentially the same argument as in Belnap 1960, p. 26, concerning Ackermann's Π'; it only needs to be shown that R satisfies the matrices.) But this means that A contains every propositional variable of R, which means that if R is formulated with an infinite number of propositional variables, then there is no such formula A. So in general $R_{/\leftrightarrow}$ has no identity.

However, it may be shown that if R is formulated with but a finite number of propositional variables $p_1, \ldots, p_m$ (let us call such a formulation R_m), then it has a formula which acts as an identity, namely, $(p_1 \rightarrow p_1) \& \cdots \& (p_m \rightarrow p_m)$ (let us call this formula t_m). The proof depends upon an application of Anderson and Belnap's 1959 Sublemma 2 (which they proved for Ackermann's Π', but which may be proven for R by simply replacing 'Π'' with 'R' wherever it occurs throughout their proof). In the course of proving Sublemma 2 they show that if a list of propositional variables $p_1, \ldots, p_m$ contains all of the propositional variables of a formula A, then $t_m \rightarrow (A \rightarrow A)$ is a theorem. But then by permutation, we have as a theorem $A \rightarrow (t_m \rightarrow A)$ which means $|A| \leq |t_m| \rightarrow |A|$, which means by P5 that $|A| \circ |t_m| \leq |A|$. The other half of the equality follows easily from the fact that $t_m \rightarrow A \rightarrow A$ is a theorem of R (it may be proven from $t_m \rightarrow A \rightarrow .t_m \rightarrow A$ *via* a permutation and a use of *modus ponens* with t_m as the $^{147}\|_{148}$ minor premise; cf. the proof of $A \rightarrow A \rightarrow B \rightarrow B$ after Theorem 1). In particular then, $t_m \rightarrow \overline{A} \rightarrow \overline{A}$ is a theorem. But by contraposition, $A \rightarrow \overline{t_m \rightarrow \overline{A}}$ is a theorem, i.e., by definition of '$\circ$', $A \rightarrow (t_m \circ A)$ is a theorem, which means that $|A| \leq |t_m| \circ |A|$, which means that $|t_m|$ is the identity of $R_{m/\leftrightarrow}$.

Although R does not in general have a formula that acts as an identity, the system R^t may be obtained as an extension of R formed by enriching the morphology with a constant proposition t and adding two axioms that ensure that t has just the properties needed to make $|t|$ the identity, namely A17, $A \rightarrow (t \rightarrow A)$, and A18, $t \rightarrow A \rightarrow A$.[8]

---

[8]These axioms give t the effect of the negation of Ackermann's "*das Absurde*," a propositional constant which he adds to his system Π' to obtain a theory of modalities,

It may be shown that R^t is a conservative extension of R in the sense that where A is a formula of R^t that does not contain t, then A is provable in R^t iff A is provable in R. The argument is a simple modification of Anderson and Belnap 1959, which showed the analogous fact that Ackermann's Π'' is a conservative extension of his Π'. We observe that if A is a t-free formula containing the propositional variables $p_1, \ldots, p_m$, then if there is a proof of A, then there is a proof of A that $\|_{149}$ contains only the propositional variables $p_1, \ldots, p_m$ (for the original proof may be rewritten, if need be, substituting one of the p_i that occur in A throughout for all of the propositional variables that do not occur in A; a simple inductive argument shows that the result is a proof of A).[9] The idea of the Anderson–Belnap proof then is to substitute t_m ($= (p_1 \to p_1)$ & $\cdots$ & $(p_m \to p_m)$) for t wherever it occurs in the proof of A. In virtue of the fact that we have already observed that t_m has the defining properties of t among the formulas that just contain the propositional variables $p_1, \ldots, p_m$, a simple inductive argument shows that the result of the substitution is a proof of A in R.

Let us record the following algebraic version of this result, the proof of which is trivial.

Theorem 2. *The Lindenbaum algebra of $R_{/\leftrightarrow}$ is isomorphically embeddable in the Lindenbaum algebra of $R^t_{/\leftrightarrow}$ under the mapping which sends $|A|$ in $R_{/\leftrightarrow}$ into $|A|$ in $R^t_{/\leftrightarrow}$.*

Both the logical and the algebraic versions of this result establish that we can study R and its Lindenbaum algebra by studying R^t and its Lindenbaum algebra. We $\|_{150}$ shall find that it is profitable to do this since R^t and its Lindenbaum algebra are more amenable to algebraic treatment.

4. R^t and De Morgan Semi-Groups with Identity

We start this section by stating a theorem, proven but not recorded in the last section.

Theorem 1. *The Lindenbaum algebra of R^t, $R^t_{/\leftrightarrow}$, is a De Morgan (indeed, an intensional) semi-group with identity $|t|$ (where the semi-group operation is defined as in Theorem 3.1).*

---

defining UA (A is impossible) as $A \to \bar{t}$. Of course, adding *das Absurde* to R does not give rise to a theory of modalities, for $A \to \bar{t}$ is simply equivalent to $\overline{A}$ (for this reason, Belnap has suggested that $\bar{t}$ when added to R be called, much more reasonably, "*das Falsche*").

[9]This step is not actually needed in the Anderson–Belnap proof, but we have been forced to include it because of the way we cast our form of their Sublemma 2 to bring out the algebraic meaning in it.

Although $R_{/\leftrightarrow}$ turned out to be a De Morgan semi-group but *not* a free one, $R^t_{/\leftrightarrow}$ is a free De Morgan semi-group with identity. (Indeed, it is a free intensional semi-group with identity; so just as free De Morgan lattices and free icdl's turned out to be the same in Chapter VIII, free De Morgan semi-groups with identity turn out to be the same as free intensional semi-groups with identity.)

We shall prove this by first defining and proving that R^t is consistent in the class of De Morgan semi-groups with identity. We accordingly say that Q is a *De Morgan semi-group with identity model* (henceforth, *model*) iff $Q = ((S, \circ, \leq, N, e), s)$, where $(S, \circ, \leq, N, e)$ (henceforth S) is a De Morgan semi-group with identity, and where s is an *assignment function* for S, i.e., a $^{150}\|_{151}$ function such that for each formula A, if A is a propositional variable, $s(A) \in S$, and if A is t, $s(A) = e$.[10]

Given a model $Q = (S, s)$, we define a *valuation* determined by Q as a function v_Q defined over all formulas and having values in S as follows: for all formulas A,

if A is a propositional variable or t, $v_Q(A) = s(A)$;

if A has the form $\overline{B}$, $v_Q(A) = Nv_Q(B)$;

if A has the form $B \vee C$, $v_Q(A) = v_Q(B)) \vee v_Q(C)$;

if A has the form $B \to C$, $v_Q(A) = v_Q(C) : v_Q(B)$.

Note that these conditions determine that

if A has the form $B \mathbin{\&} C$, $v_Q(A) = v_Q(B) \wedge v_Q(C)$;

if A has the form $B \circ C$, $v_Q(A) = v_Q(B) \circ v_Q(C)$.

Let us then say that a formula A is *true in a model* $Q = (S, s)$ iff $v_Q(A) \geq e$ (this definition is motivated by the fact that A is a theorem of R^t iff $|t| \leq |A|$, which is an immediate consequence of A18 and A19). Otherwise A is *false in the model*.

We then define a formula as *valid in a De Morgan semi-group with identity* S iff it is true in every model $Q = (S, s)$, and otherwise as *falsifiable in a De Morgan semi-group with identity* S. $^{151}\|_{152}$

We finally define a formula as *valid in a class of De Morgan semi-groups with identity* iff it is valid in each member of the class, and otherwise as *falsifiable in the class*. If the particular class in question is the class of all

---

[10]Note that this definition, as well as the subsequent definitions, is applicable to R as well as R^t, the "*if A is t*" clause being vacuous in application to R.

De Morgan semi-groups with identity, then we shall say simply that the formula is *valid* or *falsifiable*.

We shall say that R or R^t is *consistent with respect to a class of De Morgan semi-groups with identity* iff, if a formula A is provable, then it is valid in that class. We shall say that R or R^t is *complete with respect to a class of De Morgan semi-groups with identity* iff, if a formula A is valid in the class, then it is provable. We shall call R or R^t just *consistent* or *complete* iff the particular class in question is the class of all De Morgan semi-groups with identity.

We may now prove

Theorem 2. *Both R and R^t are consistent; hence both R and R^t are consistent in the class of intensional semi-groups with identity.*

Since R is contained in R^t, and since intensional semi-groups are De Morgan semi-groups, it suffices to show R^t consistent. We show that the axioms are valid and that the rules preserve validity. Since all of the axioms have the form of implications, we may show an axiom $A \to B$ valid by showing that for every valuation v_Q, $v_Q(A) \le v_Q(B)$, for then by P16, $e \le v_Q(B) : v_Q(A)$. The validity of A1, A7–A9, A11–A12, and A14 then follows immediately $^{152}\|_{153}$ from De Morgan lattice properties, since (schematically) these axioms contain the arrow only as the principal connective. As an illustration, we consider A11: $A \to A \vee B$. By inductive definition of v_Q, $v_Q(A \vee B) = v_Q(A) \vee v_Q(B)$. But $v_Q(A) \le v_Q(A) \vee v_Q(B)$ is a lattice property. The validity of the remaining axioms follows trivially from the properties of De Morgan semi-groups with identity that we developed in Section 2. Thus A2′ is treated by P6, A3 by P4, A4 by P18, A5 by P20, A6 by P19, A10 by P10, A13 by P11, and A17 and A18 by P14. As an illustration, we shall demonstrate the validity of A2′: $A \to (B \to C) \to$ $.B \to (A \to C)$. By inductive definition of v_Q, $v_Q(A \to (B \to C)) = (v_Q(C) : v_Q(B)) : v_Q(A)$, and $v_Q(B \to (A \to C)) = (v_Q(C) : v_Q(A)) : v_Q(B)$, but by P6 these are identical.

We next show that the rules preserve validity. For *modus ponens*, suppose $e \le v_Q(A)$ and $e \le v_Q(A \to B)$. But then $e \le v_Q(B) : v_Q(A)$, and by P14, $v_Q(A) \le v_Q(B)$. But then by transitivity, $e \le v_Q(B)$. For adjunction, we assume that $e \le v_Q(A)$ and $e \le v_Q(B)$; but then $e \le v_Q(A) \wedge v_Q(B) = v_Q(A \,\&\, B)$.

From Theorem 2, we obtain

Theorem 3. *The Lindenbaum algebra of R^t formulated with n propositional variables is the free De Morgan semi-group with identity with n free generators; and it is also the free intensional semi-group with identity with n free generators.*

This theorem is proven much like Theorem VIII:2. We prove the first part by considering any formulation of $^{153}\|_{154}$ R^t with n propositional variables p_i. Clearly the $|p_i|$ generate $\mathrm{R}^t_{/\leftrightarrow}$. Now consider any mapping f of these generators into an arbitrary De Morgan semi-group with identity S. Now define an assignment function s for S so that $s(p_i) = f(|p_i|)$ and $s(t) = e$. We thus have a model $Q = (S, s)$ and its valuation v_Q. Define $h(|A|) = v_Q(A)$. It is easy to see that if h is a well-defined (single-valued) mapping then it is an e-homomorphism. But that h is well defined follows from Theorem 2. For suppose $|A| = |B|$, i.e., that both $A \to B$ and $B \to A$ are theorems of R^t. But then by Theorem 2, $e \le v_Q(A \to B) = v_Q(B) : v_Q(A)$ and $e \le v_Q(B \to A) = v_Q(A) : v_Q(B)$. But then by P16, $v_Q(A) = v_Q(B)$, and $h(|A|) = h(|B|)$. The second part of the theorem follows immediately from the fact that $\mathrm{R}^t_{/\leftrightarrow}$ is an intensional semi-group with identity (Theorem 1) (cf. proof of second part of Theorem VIII:2).

We remark that Theorem 3 identifies (up to e-isomorphism) the free De Morgan semi-group with identity with n free generators and the free intensional semi-group with identity with n free generators, and that it is an immediate consequence of this fact that every De Morgan semi-group with identity is an e-homomorphic image of an intensional semi-group with identity (cf. proof of Theorem VIII:3).

We next observe that it is a consequence of the fact that $\mathrm{R}^t_{/\leftrightarrow}$ is an intensional semi-group with identity that $^{154}\|_{155}$

Theorem 4. R^t *is complete with respect to the class of intensional semi-groups with identity; hence, R^t is complete.*

The second part of the theorem follows from the first in virtue of the fact that an intensional semi-group with identity is a De Morgan semi-group with identity. We prove the first part by showing that if a formula A is not provable in R^t, then it is falsifiable in $\mathrm{R}^t_{/\leftrightarrow}$. We define the *canonical model* $C = (\mathrm{R}^t_{/\leftrightarrow}, c)$, where c is the *canonical assignment function*, i.e., if A is a propositional variable or t, $c(A) = |A|$. We next define the *canonical valuation* v_C as the valuation determined by C. It is easy to see that $v_C(A) = |A|$. We next observe that if A is not provable in R^t, then $|t| \not\le |A|$. This is because otherwise $t \to A$ would be provable, and since t is also provable,[11] A would be provable by *modus ponens*. Since $|t|$ is the identity of $\mathrm{R}^t_{/\leftrightarrow}$, this completes the theorem.

Putting Theorems 2 and 4 together gives us

Theorem 5. *A formula A is provable in R^t iff it is valid, and iff it is valid in the class of intensional semi-groups with identity.*

---

[11]It follows from $t \to t \to t$ (instance of A18) and $t \to t$ (instance of A1) by *modus ponens*.

Even though R does not have a formula like t that acts like an identity, we get by virtue of the embedding $^{155}\|_{156}$ of $R_{/\leftrightarrow}$ into $R^t_{/\leftrightarrow}$ of Theorem 3.2, as an immediate consequence of Theorem 4,

Theorem 6. *R is complete with respect to the class of intensional semi-groups with identity; hence, R is complete.*

Combining Theorem 6 and Theorem 2, we get

Theorem 7. *A formula A is provable in R iff it is valid, and iff it is valid in the class of intensional semi-groups with identity.*

5. An Algebraic Analogue to the Admissibility of Ackermann's Rule γ

Ackermann's system Π', which provided a stimulus for Anderson and Belnap's formulation of their system E, contained the following primitive rule:

(γ) From A and $\overline{A} \vee B$, infer B.

Neither the system E nor the system R contain γ as a primitive rule, but Anderson and Belnap conjecture that γ is *admissible* for both systems in the sense that whenever A is a theorem and $\overline{A} \vee B$ is a theorem, then B is a theorem.[12] These conjectures constitute two of the major open problems of intensional logic. $^{156}\|_{157}$

To grasp the importance of γ is to see that its admissibility for E and R is a necessary condition for the systems' having a *normal* semantics, in a sense to be explained. Following Church 1953, let us say of a model for a propositional calculus containing signs for disjunction ($\vee$) and negation ($^-$) that it is *normal* iff i) it contains a certain subset of *designated* elements, and ii) valuations in it are such that $\overline{A}$ receives a designated value just when A receives an undesignated value, and $A \vee B$ receives a designated value just when either A does or B does. Let us then say of a class of normal models that it is *characteristic* for a propositional calculus iff a formula is a theorem just when it receives a designated value under every valuation in a member of the class. It may be seen immediately that the admissibility of γ is a necessary condition for any propositional calculus's having a *normal* semantics in the sense of its having a characteristic class of normal models.

It is natural to extend Anderson and Belnap's conjectures concerning the admissibility of γ to the system R^t. This conjecture, expressed in terms of the Lindenbaum algebra of R^t, takes the following form:

---

[12]Cf. Anderson 1963 for the published conjecture concerning E. The conjecture concerning R was communicated personally. There are good reasons for not having γ primitive, which are stated in Anderson's paper, as well as in Belnap 1959.

(I) $|t| \le |A|$ and $|t| \le N|A| \vee |B|$ implies $|t| \le |B|$.

Indeed, (I) suggests a conjecture concerning De Morgan semi-groups with identity in general, that is an obvious algebraic analogue of the admissibility of γ, namely, $^{157}\|_{158}$

(I') $e \le a$ and $e \le Na \vee b$ implies $e \le b$.

We shall find at the end of this section that such a conjecture would be false, but we postpone this denouement until we first develop some interesting equivalents of property (I').

We first observe that (I') is equivalent to a simpler property

(II) $e \le Ne \vee b$ implies $e \le b$.

Property (II) is obviously only a special case of (I'), substituting e for a. To see that (II) implies (I'), we assume (II) and the hypotheses of (I'). The hypotheses of (I') easily imply $e \le a \wedge (Na \vee b)$, which in turn implies *via* a distribution $e \le (a \wedge Na) \vee (a \wedge b)$. By P17, $a \wedge Na \le a \circ Na$. By P15, $e \le a{:}a$, which means that $N(a{:}a) \le Ne$. But by Theorem 1.1, $N(a{:}a) = a \circ Na$, so $a \wedge Na \le Ne$, which means that $e \le Ne \vee (a \wedge b)$. But then by (II), $e \le a \wedge b$, and so $e \le b$, which completes the equivalence.

We next show that (II) is equivalent to

(III) $b \not\le Ne$ implies there exists a truth filter T such that $b \in T$ but $Ne \notin T$.

It is easy to see that (II) is a necessary condition for (III), for suppose that $e \le Ne \vee b$, but $e \not\le b$. Then $Nb \not\le Ne$, and by (III) there exists a truth filter T such that $Nb \in T$ but $Ne \notin T$. But since $Ne \notin T$, $e \in T$, and hence $Ne \vee b \in T$. But since truth filters are prime, then either $Ne \in T$ or $b \in T$. But we know that $Ne \notin T$, and since $Nb \in T$, then $b \notin T$ as well. But this is a contradiction, so contrary to hypothesis, $e \le b$. $^{158}\|_{159}$

The sufficiency of (II) is a bit harder. We suppose that $b \not\le Ne$ and consider the filter generated by b together with e, $F(b, e)$. It is easy to see that $F(b, e)$ is consistent in the sense that it contains for no element a, both a and Na. If it did, then $e \wedge b \le a \wedge Na$, and since $a \wedge Na \le Ne$ (cf. the argument that showed the equivalence of (I') and (II)), then $e \wedge b \le Ne$. But then $e \le Ne \vee Nb$, and by (II), $e \le Nb$, i.e., $b \le Ne$, contrary to the hypothesis of (III).

Now that we know there is at least one consistent filter containing both b and e, we consider the set of all such consistent filters, ordered by set inclusion. It is easy to see that the union of any chain of consistent filters containing both b and e is itself a consistent filter containing both b and e.

So every chain has an upper bound, and we have the hypothesis of Zorn's lemma, so we may conclude that there is a consistent filter T that contains both b and e and is maximal with respect to these properties. We now show that T is exhaustive in the sense that it contains for every element a, at least one of a and Na.

Suppose that T contained for some element a, neither a nor Na. Then consider the filter generated by T together with a, $F(T, a)$, and the filter generated by T together with Na, $F(T, Na)$. Since these are filters which properly include T, and since T was shown maximal with respect to being a consistent filter containing both b and e, then these filters must be inconsistent. But this means that [159] for some elements $t_1, t_2 \in T$, and for some elements c and d,

$$t_1 \wedge a \le c \wedge Nc, \text{ and } t_2 \wedge Na \le d \wedge Nd.$$

But since for any element x, $x \wedge Nx \le Ne$ (cf. the proof of the equivalence of (I′) and (II)), then

$$t_1 \wedge a \le Ne \text{ and } t_2 \wedge Na \le Ne.$$

Then setting $t = t_1 \wedge t_2$ we have for some element $t \in T$,

$$t \wedge a \le Ne \text{ and } t \wedge Na \le Ne.$$

But then we have

$$(t \wedge a) \vee (t \wedge Na) \le Ne,$$

which leads by distribution to

$$t \wedge (a \vee Na) \le Ne.$$

But since $e \le a \vee Na$ (which follows from $a \wedge Na \le Ne$), then $t \wedge (a \vee Na) \in T$, and hence $Ne \in T$. But this is a contradiction since T was shown consistent. So T is both consistent and exhaustive and hence a truth filter, which completes the proof.

It is interesting to observe that (III) is a generalization of Stone's 1936 theorem concerning maximal filters in Boolean algebras. Stone proved that for any element $b \ne 0$, there exists a maximal (proper) filter M such that $b \in M$. The generalization is obvious once it is recalled that a Boolean algebra is a De Morgan semi-group with identity, and that a filter is maximal in a Boolean algebra iff it contains for every element, exactly one of a and $\bar{a}$. Indeed, our proof that (II) implies (III) was essentially based upon Stone's proof, for (II) is trivially true of Boolean algebras. [160]

Now let us at last discuss the truth of (I'). It is an easy consequence of the equivalence of (I') with (III) that (I') does not hold of any De Morgan semi-group with identity whose underlying De Morgan lattice does not have a truth filter, i.e., whose underlying De Morgan lattice is not an icdl. One might next conjecture that perhaps (I') holds of intensional semi-groups with identity, but this conjecture too would be false. For a counter-example, consider the intensional semi-group with identity whose underlying icdl is the sub-icdl of M_0 generated by $\{+1, +0, +3\}$, and whose multiplication is defined by the following matrix:

$\circ$	-3	-1	-0	$+0$	$+1$	$+3$
-3	-3	-3	-3	-3	-3	-3
-1	-3	-1	-0	-3	-1	-0
-0	-3	-0	-0	-3	-0	-0
$+0$	-3	-3	-3	$+0$	$+0$	$+0$
$+1$	-3	-1	-0	$+0$	$+1$	$+3$
$+3$	-3	-0	-0	$+0$	$+3$	$+3$

It may be routinely, though laboriously, verified that this is indeed an intensional semi-group with identity, and indeed it may be shown that on this underlying icdl it is the only intensional semi-group whose identity is $+1$.

We now observe that property (II) (and at the same time property (I')) fails in it, since $+1 \leq -1 \vee +0$, and yet $+1 \nleq +0$. It is interesting to note that the underlying icdl $^{161}\|_{162}$ served as an earlier counter-example to a γ-like property of icdl's. Belnap and Spencer 1964 showed that it is not true of every icdl that any consistent filter is extendable to a consistent, exhaustive filter (unlike the case with Boolean algebras), and they did this by observing that the filter generated by $+1$ in this icdl is consistent, but not so extendable. What we have done with our present counter-example is to introduce a De Morgan multiplication upon this icdl so as to make $+1$ the identity.

6. The Algebraic Structure of E

We may define a multiplication operation upon the Lindenbaum algebra of E, as we did upon $R_{/\leftrightarrow}$, but the result is not so pleasant since the multiplication then turns out not to be associative. Associativity is equivalent to permutation, but it is exactly the addition of the permutation axiom A2' to E which gives R. These remarks do not merely reflect a prejudice against non-associative algebraic systems, but are based upon the repeated failure to find perspicuous axioms involving multiplication that capture the restricted permutation axiom A2* of E.[13] [162]||[163]

However, Robert K. Meyer 1966 has made a plausible conjecture which, if true, offers a graceful way out. He offers for consideration a logical system R^L, gotten from R by adding a necessity operator L and the following axioms and rules, characteristic of S4.[14]

LA1. $L(A \to B) \to .LA \to LB$

LA2. $LA \to A$

LA3. $LA \to LLA$

LA4. $(LA \,\&\, LB) \to L(A \,\&\, B)$

LR5. If A is an axiom, so is LA.

Meyer then conjectures that the arrow of E can be defined in terms of the arrow of R by $A \to_E B = L(A \to_{R^L} B)$ so that a formula is provable in E iff its obvious definitional transform is provable in R^L. Meyer points out that this definition is analogous to Lewis's definition of strict implication in terms of necessary material implication.

All of this suggests that just as McKinsey and Tarski studied the algebra of strict implication in S4 by considering Boolean algebras with a Kuratowski closure operation, so we may study the algebra of entailment by [163]||[164] studying De Morgan semi-groups with an appropriate closure operation. Indeed, since it is easy to prove that t and its axioms can be added conservatively to R^L,[15] we may study the algebra of the resulting system R^{Lt}.

Let us then define a *De Morgan closure operation* C upon a De Morgan semi-group with identity as a unary operation satisfying (where Ia is defined as $NCNa$[16])

C1. $a \leq Ca$

C2. $CCa = a$

---

[13]Consider as an example of an unperspicuous axiom, $N(a_1 \circ a_2) \circ (b \circ N(c_1 \circ c_2)) = (N(a_1 \circ a_2) \circ b) \circ N(c_1 \circ c_2)$.

[14]Indeed, LA1–LA3 are the axioms added to the classical propositional calculus in the Gödel formulation of S4, and LR5 is but a special case of Gödel's rule (if A is a theorem, so is LA), which Meyer shows holds in full generality for R. Axiom LA4 would be redundant in the Gödel formulation of S4, but it is not in the Meyer formulation of R^L, for reasons special to R.

[15]In the sense of Section 3 when it was shown that R^t is a *conservative* extension of R.

[16]The operation I may be called a *De Morgan interior operation*, and is dual to C.

C3. $C(a \vee b) = Ca \vee Cb$

C4. $C(Ne) = Ne$

C5. $Ca \circ Ia \leq C(a \circ b)$,

and call the resulting structure a *closure De Morgan semi-group with identity*. We remark that axiom C4 is easily shown to be equivalent to

C4'. $I(a : b) \leq Ia : Ib$.

The operation C is a specialization of a very general notion of a *closure operation* on a lattice[17] defined by C1, C2, and

C6. $a \leq b$ implies $Ca \leq Cb$,

for it is easily shown that C6 follows from C3. We remark [164]‖[165] further that C has marked similarities to the *Kuratowski closure operation* of McKinsey and Tarski 1944, which is defined upon a Boolean algebra by C1, C2, C3, and

C7. $C(0) = 0$.

Indeed, when a Boolean algebra is considered as a De Morgan semi-group with identity in the way described in Section 2, then $\wedge$ becomes $\circ$ and 1 becomes e, and C7 becomes a special case of C4, and C5 becomes derivable. So a closure De Morgan semi-group with identity may be looked at as a generalization of a closure Boolean algebra (which seems to add strength to Meyer's conjecture).

To make a long story short, upon defining MA (A is possible) in R^{Lt} as $\overline{L(\overline{A})}$, and then defining on the Lindenbaum algebra of R^{Lt}, $R_{/\leftrightarrow}^{Lt}$,

$$C|A| = |MA|,$$

it may be easily shown that the resulting structure is a closure De Morgan semi-group with identity. We may then restrict our notion of a model in Section 4 appropriately and add to our inductive definition of a valuation:

if A is of the form LB, $v_Q(A) = Iv_Q(B)$, i.e.,

if A is of the form MB, $v_Q(A) = Cv_Q(B)$.

---

[17]Cf. esp. Ward 1942, p. 192; Certaine 1944, p. 22; and Birkhoff 1948, p. 49.

It is then easy to prove that R^{Lt}, and hence R^L, are appropriately consistent and complete in the class of closure De Morgan semi-groups with identity, and hence that $R^{Lt}_{/\leftrightarrow}$ is appropriately free in the class. The only trick is to see that the effect of Gödel's rule is gotten by showing that $e \leq a$ implies $e \leq Ia$, which follows $^{165}\|_{166}$ easily from C6 and C4.

To make a short story even shorter, everything that was done in the previous sections for R^t and R in their relation to De Morgan semi-groups with identity can be mimicked for R^{Lt} and R^L in their relation to closure De Morgan semi-groups with identity—time, space, and patience permitting.

7. Closed and Open Problems

Like the closed and open sets of a topology, our closed and open problems are intimately connected. But just as one may talk of a topology in terms of both its closed and its open sets, so we may discuss our study in terms of both its closed and its open problems. We shall talk first of its closed problems.

We have developed a representation theory for icdl's and De Morgan lattices in Chapters II, III, and IV that is parallel to the representation theory for Boolean algebras developed by Stone. Then in Chapters V and VI we have made connections between our representation theory for icdl' s and De Morgan lattices and the Boolean representation theory. In Chapters VII and VIII we have connected the system B of first degree entailments of the intensional logics E and R to De Morgan lattices and icdl's, giving at the same time an algebraic proof of the completeness of B. These results have parallels to $^{166}\|_{167}$ well-known connections between the classical propositional calculus and Boolean algebras. In Chapter IX, we have tried to give an intuitive description of what De Morgan lattices are all "about," showing that every De Morgan lattice can be interpreted ultimately in terms of "topics." This leads by the completeness results of Chapters VII and VIII to an intuitive completeness proof for the system B. Finally, in Chapter X we have made connections between R and a special kind of residuated semi-group, and have suggested how these connections could be extended to E.

But most of these results suggest further open problems. Among these open problems are the following:

1) The proofs and theorems of Chapter II for icdl's and of Chapter III for De Morgan lattices have clear similarities. Find an interesting connection between the two bodies of results.

2) Find an algebraic completeness proof for the first degree entailment fragments of EQ and RQ. In particular, try to complete the incomplete proof sketched in Section 6, Chapter VII. It might also prove fruitful to try to develop a theory of "intensional polyadic algebras" akin to Halmos's

1962 theory of "polyadic algebras" that have proved so useful in the algebraic study of the classical predicate calculus. Or alternatively, one might develop a way of treating EQ and RQ by adding axioms for identity and creating an appropriate theory akin to Tarski and Thompson's 1952 theory of "cylindrical algebras." [167]‖[168]

3) Give an interesting algebraic proof for the completeness of the first degree formula fragments of E and R. Cf. Section 5, Chapter I.

4) Make the "intuitive" semantics of Chapter IX even more intuitive. Perhaps this could be done by developing a non-classical (intensional) information theory. Also extend this intuitive semantics beyond the first degree entailments.

5) Circumscribe the class of intensional semi-groups with identity having the γ-property (I′) of Section 5, Chapter X. It would be desirable to do so in such a way that the Lindenbaum algebra of R^t falls in the class, thus solving the problem of the admissibility of γ for R.

6) Develop a representation theory for intensional and De Morgan semi-groups with identity. In particular, it would be nice to be able to show that any finite sub-icdl of an intensional semi-group with identity can have an operation of residuation defined upon it in such a way as to make it an intensional semi-group with the same identity as the original and such that the new operation of residuation gives the same results as the old when these results happen to lie in the sub-icdl. This, in connection with Theorem 5 of Chapter VIII, should give a decision procedure for R akin to the decision procedure of McKinsey 1941 for S4. Cf. also Rasiowa and Sikorski 1963, pp. 101 and 478. [168]‖[169]

NOTATION

Notation	*Interpretation*
Theorem III:2.1, etc.	Theorem 1 of Section 2 of Chapter III, etc. (If the citation is in the same chapter in which the theorem is proven then the chapter reference is omitted, and similarly for sections.)
iff	if, and only if
$x \in A$, $x \notin A$	x is, or is not, a member of A
$\{x: \ldots x \ldots\}$	the set of x such that $\ldots x \ldots$
$A \subseteq B$	A is a subset of B
$A \cap B$	the intersection of A and B
$A \cup B$	the union of A and B
$\overline{A}$	the (set-theoretical) complement of A
εA	A or $\overline{A}$, p. 27
$\emptyset$	the empty set
$x < c$, $x \not< c$	the cardinal x is, or is not, (strictly) less than the cardinal c. (We take the von Neumann definition of a cardinal so that $x < c$ iff $x \in c$.)
$\aleph_0$	the cardinality of the natural numbers
$(a_1, a_2, \ldots, a_n)$	an n-tuple
$\{a_x\}_{x \in X}$, $\{a_x\}_{x < c}$	an indexed set, i.e., a mapping which assigns to every $x \in X$, or $x < c$, an element a_x
$\times_{x \in X} A_x$, $\times_{x < c} A_x$	Cartesian product

103

$a \prec b$	a (strictly) precedes b in the well-order, p. 35
$a = b$, $a \neq b$	a is, or is not, identical with b
$a \equiv b$, $a \not\equiv b$	a is, or is not, equivalent (often, congruent) to b, p. 4, p. 41, p. 43 [169] $\|_{170}$
$\|a\|$	$\{x : x \equiv a\}$, p. 5
$a \leq b$, $a \not\leq b$	a is, or is not, less than or equal to b, p. 6
$a < b$, $a \not< b$	$a \leq b$ but $a \neq b$, it is not the case that $a < b$
$a \wedge b$	the meet of a and b, p. 6
$\bigwedge A$, $\bigwedge_{x \in X} a_x$	the (generalized) meet of A, or $\{a_x\}_{x \in X}$, p. 6
$a \vee b$	the join of a and b, p. 6
$\bigvee A$, $\bigvee_{x \in X} a_x$	the (generalized) join of A, or $\{a_x\}_{x \in X}$, p. 6
$\bar{a}$	the (Boolean) complement of a, p. 9
$\mathrm{N}a$	the De Morgan, or intensional complement of a, p. 23, p. 21
a^{-1}	group inverse of a, p. 82
$-a$	the pseudo-complement of a, p. 11
$a \Rightarrow b$	the relative pseudo complement, p. 10
$a \supset b$	$\bar{a} \vee b$
$a \circ b$	semi-group multiplication, p. 82, usually of a De Morgan semi-group, p. 85
$a : b$, $a :: b$	right and left residuals, p. 82
$a + b$	$\mathrm{N}(\mathrm{N}a \circ \mathrm{N}b)$ in a De Morgan semi-group, p. 85
$\dot{-}$, $\ddot{-}$	right and left differences, p. 85
$\mathrm{C}a$	the closure of a, p. 10, p. 98
$\mathrm{I}a$	the interior of a, p. 11, p. 98
F_a	the filter generated by a, p. 27, p. 8

0	the least element, p. 9
1	the greatest element, p. 9
e	semi-group identity, p. 82 [170]‖[171]
1	the degenerate one element lattice, p. 6
2	the two element lattice, p. 9
M_0	p. 22
D	p. 37
$\prod_{x \in X} A_x,\ \prod_{x < c} A_x$	direct product (usually of De Morgan lattices) or product of icdl's w/t-f p. 25
M^c	product of M_0, p. 29
D^c	direct product of D, p. 39
B	System of First Degree Entailments, p. 69
E	System of Entailment, p. 1
R	System of Relevant Implication, p. 1
R^t	R with a constant strongest theorem t, p. 89
R^L, R^{Lt}	R, or R^t, with necessity, p. 98, p. 98
$A \,\&\, B$	truth-functional conjunction, p. 1, p. 13
$A \vee B$	truth-functional disjunction, p. 1, p. 13
$\overline{A}$	truth-functional negation, p. 1, p. 13
$A \supset B$	material implication, p. 13
$A \rightarrow B$	A entails B (when a formula of E), or A relevantly implies B (when a formula of R), though we often read both as the former, and also call them both entailments, p. 1
$A \leftrightarrow B$	$(A \rightarrow B) \,\&\, (B \rightarrow A)$, p. 62
$A \circ B$	A is consistent with B, p. 87
LA	A is necessary, p. 98

MA	A is possible, p. 99 [171]$\|_{172}$
$A \equiv B$	$A \supset B$ and $B \supset A$ are both theorems, p. 14
$|A|$	set of formulas that (provably) co-imply, or co-entail A, p. 14, p. 62, p. 69
$B_{/\leftrightarrow}$	Lindenbaum algebra of B, p. 70
$E_{/\leftrightarrow}$	Lindenbaum algebra of E, p. 63
$R_{/\leftrightarrow}$, $R^t_{/\leftrightarrow}$, $R^{Lt}_{/\leftrightarrow}$	Lindenbaum algebra of R, p. 63, of R^t p. 90, of R^{Lt}, p. 99
t	constant least theorem, p. 89
T	truth filter, p. 21
v_Q	valuation determined by the model Q, p. 61, p. 91 [172]$\|_{173}$

BIBLIOGRAPHY

ACKERMANN, WILHELM

1956   "Begründung einer strengen Implikation," *Journal of Symbolic Logic*, vol. 21, pp. 113–128.

ANDERSON, ALAN ROSS

1963   "Some open problems concerning the system E of entailment," *Acta Philosophica Fennica*, vol. 23, pp. 7–18.

ANDERSON, ALAN ROSS, AND BELNAP, NUEL D., JR.

1959   "Modalities in Ackermann's 'rigorous implication'," *Journal of Symbolic Logic*, vol. 24, pp. 107–111.

1962a   "Tautological entailments," *Philosophical Studies*, vol. 13, pp. 9–24.

1962b   "The pure calculus of entailment," *Journal of Symbolic Logic*, vol. 27, pp. 19–52.

1963   "First degree entailments," *Mathematische Annalen*, vol. 149, pp. 302–319.

1966   *Entailment*, mimeographed material for a forthcoming book.

ANGELL, R. B.

1962   "A propositional logic with subjunctive conditionals," *Journal of Symbolic Logic*, vol. 27, pp. 327–343.

BELNAP, NUEL D., JR.

1960   "A formal analysis of entailment," Technical Report No. 7, Office of Naval Research, Group Psychology Branch, Contract SAR/Nonr-609(16), New Haven, 1960.

1965   "Intensional models for first degree formulas," mimeographed, and forthcoming in *Journal of Symbolic Logic*.

BELNAP, NUEL D., JR., AND SPENCER, J. H.

1964   "Intensionally complemented distributive lattices," mimeographed.

BIAŁYNICKI-BIRULA, A., AND RASIOWA, H.

1957   "On the representation of quasi-Boolean algebras," *Bulletin de l'Academie Polonaise des Sciences*, vol. 5, pp. 259–261.

BIRKHOFF, GARRETT

1933   "On the combination of subalgebras," *Proceedings of the Cambridge Philosophical Society*, vol. 29, pp. 441–464.

1935   "On the structure of abstract algebras," *Proceedings of the Cambridge Philosophical Society*, vol. 31, pp. 433–454.

1942   "Lattice-ordered groups," *Annals of Mathematics*, vol. 43, pp. 298–330.

1944   "Subdirect unions in universal algebras," *Bulletin of the American Mathematical Society*, vol. 50, pp. 764–768.

1948   *Lattice Theory*, Providence, R.I.

BIRKHOFF, GARRETT, AND VON NEUMANN, J.

1936   "The logic of quantum mechanics," *Annals of Mathematics*, vol. 37, pp. 823–843.

BOOLE, GEORGE

1847   *The Mathematical Analysis of Logic*, Cambridge.

CARNAP, RUDOLF, AND BAR-HILLEL, YEHOSHUA

1952   "An outline of a theory of semantic information," M.I.T. Research Lab. of Electronics Technical Report No. 247.

CERTAINE, J.

1943   "Lattice-ordered groupoids and some related problems," Harvard Doctoral Thesis.

CHURCH, ALONZO

1951   "The weak theory of implication," *Kontrolliertes Denken*, ed. by A. Menne, A. Wilhelmy, and Helmut Angsil, Munich, pp. 22–37.

1956   *Introduction to Mathematical Logic*, 2nd ed., vol. I, Princeton, N.J.

DUNN, J. MICHAEL, AND BELNAP, NUEL D., JR.

1965   "Homomorphisms of intensionally complemented distributive lattices," mimeographed, and presented to a meeting of the Association for Symbolic Logic, New York, December 28, 1965, (abstract forthcoming in the *Journal of Symbolic Logic*).

FITCH, FREDERIC B.

1952   *Symbolic Logic*, New York.

FUCHS, L.

1963   *Partially Ordered Algebraic Systems*, Budapest.

GOODMAN, NELSON

1961   "About," *Mind*, vol. LXX, no. 277, pp. 1–24.

HALMOS, PAUL R.

1962   *Algebraic Logic*, New York.

HENKIN, LEON

1954   "Boolean representation through propositional calculus," *Fundamenta Mathematicae*, vol. 41, pp. 89–96.

KALMAN, J. A.

1958   "Lattices with involution," *Transactions of the American Mathematical Society*, vol. 87, pp. 485–491.

KISS, STEPHEN A.

1961   *An Introduction to Algebraic Logic*, privately printed, Westport Connecticut (available at 3 Laurel Lane).

Łoś, J.

  1957   "Remarks on Henkin's paper: Boolean respresentation through propositional calculus," *Fundamenta Mathematicae*, vol. 44, pp. 82–83.

McCall, Storrs

  1965   "Connexive implication," mimeographed.

McKinsey, J. C. C.

  1941   "A solution to the decision problem for the Lewis systems S.2 and S.4 with an application to topology," *Journal of Symbolic Logic*, vol. 6, pp. 117–134.

McKinsey, J. C. C., and Tarski, Alfred

  1944   "The algebra of topology," *Annals of Mathematics*, vol. 45, pp. 141–191.

  1946   "On closed elements in closure algebras," *Annals of Mathematics*, vol. 47, pp. 122–162.

  1948   "Some theorems about the sentential calculi of Lewis and Heyting," *Journal of Symbolic Logic*, vol. 13, pp. 1–15.

Meyer, Robert K.

  1966   "Topics in modal and many valued logics," University of Pittsburgh Doctoral Thesis.

Monteiro, A.

  1960   "Matrices de Morgan caractéristiques pour le calcul propositionnel classique," *Anais da Academia Brasileira de Ciências*, vol. 32, pp. 1–7.

Prawitz, Dag

  1965   *Natural Deduction*, Uppsala.

Rasiowa, Helena and Sikorski, Roman

  1963   *The Mathematics of Metamathematics*, Warsaw.

SIKORSKI, ROMAN

   1964   *Boolean Algebras*, Berlin.

SMILEY, T. J.

   1959   "Entailment and deducibility," *Proceedings of the Aristotelian Society*, vol. 59, pp. 233–254.

STONE, M. H.

   1936   "The theory of representations for Boolean algebras," *Transactions of the American Mathematical Society*, vol. 40, pp. 37–111.

   1937   "Topological representation of distributive lattices and Brouwerian logics," *Časopis pro pěstování matematiky a fysiky*, vol. 67, pp. 1–25.

STRAWSON, P. F.

   1952   *Introduction to Logical Theory*, London.

TARSKI, ALFRED

   1935   "Grundzüge des Systemenkalkül, Erster Teil," *Fundamenta Mathematicae*, vol. 25, pp. 503–526. Translated, along with the second part, which originally appeared in *Fundamenta Mathematicae*, vol. 26 (1936), pp. 283–301, in *Logic, Semantics, Metamathematics, Papers from 1923 to 1938 by Alfred Tarski*, tr. by J. H. Woodger, Oxford, 1956.

TARSKI, ALFRED, AND THOMPSON, F. B.

   1952   "Some general properties of cylindric algebras," *Bulletin of the American Mathematical Society*, vol. 58, p. 65, abstract 85.

WARD, M.

   1942   "The closure operators of a lattice," *Annals of Mathematics*, vol. 43, pp. 191–196.

WARD, M., AND DILWORTH, R. P.

   1939   "Residuated lattices," *Transactions of the American Mathematical Society*, vol. 45, pp. 335–354.

WHITEHEAD, ALFRED NORTH, AND RUSSELL, BERTRAND

    1910   *Principia Mathematica*, vol. 1, Cambridge, England.

WHITMAN, P. M.

    1961   "Status of word problems for lattices," *Lattice Theory*, Proceedings of Symposia in Pure Mathematics, vol. II, Providence, R.I.